Get started with MicroPython on Raspberry Pi Pico, 2nd Edition

T0332326

Get started with MicroPython on Raspberry Pi Pico
by Gareth Halfacree and Ben Everard
ISBN: 978-1-912047-29-1
Copyright © 2024 Gareth Halfacree and Ben Everard
Printed in the United Kingdom
Published by Raspberry Pi Ltd., 194 Science Park, Cambridge, CB4 0AB

Editors: Brian Jepson, Liz Upton
Interior Designer: Sara Parodi
Production: Nellie McKesson
Photographer: Brian O'Halloran
Illustrator: Sam Alder
Graphics Editor: Natalie Turner
Publishing Director: Brian Jepson
Head of Design: Jack Willis
CEO: Eben Upton

June 2024: Second Edition
January 2021: First Edition

Table of Contents

Appendices

Welcome

You might think of computers as things you stick on your desk and type on. That is certainly one type of computer, but it's not the only type. In this book, we're looking at microcontrollers — small processing units with a bit of memory that are good at controlling other hardware. You probably have lots of microcontrollers in your house already.

There's a good chance your washing machine is controlled by a microcontroller; maybe your watch is; you might find one in your coffee machine or microwave. All these microcontrollers already have software running on them and the manufacturers make it hard to make any kind of change to that software.

A Raspberry Pi Pico, on the other hand, is a microcontroller that you can easily program (and reprogram!) over a USB connection. In this book, we'll look at how to get started with Pico, and how to make it work with other electronic components. By the end of the book, you'll know how to create your own programmable electronic contraptions. What you do with them is up to you.

You can find this book's example code, errata, and other resources in its GitHub repository at **rptl.io/pico-resources-2e**. If you've found what you believe is a mistake or error in the book, please let us know by using our errata submission form at **rptl.io/pico-errata-2e**.

About the authors

Gareth Halfacree is a freelance technology journalist, writer, and former system administrator in the education sector. With a passion for open-source software and hardware, he was an early adopter of the Raspberry Pi platform and has written several publications on its capabilities and flexibility. He can be found on Mastodon as **@ghalfacree@mastodon.social** or via his website at **freelance.halfacree.co.uk**.

Ben Everard is a geek who has stumbled into a career that lets him play with new hardware. As the editor of *HackSpace* magazine (**hsmag.cc**), he spends more time than he really should experimenting with the latest (and not-so latest) DIY tech. He lives in Bristol with his wife and two daughters in a house that's slowly filling up with electronics equipment and 3D printers.

Colophon

Raspberry Pi is an affordable way to do something useful, or to do something fun.

Democratising technology — providing access to tools — has been our motivation since the Raspberry Pi project began. By driving down the cost of general-purpose computing to below $5, we've opened up the ability for anybody to use computers in projects that used to require prohibitive amounts of capital. Today, with barriers to entry being removed, we see Raspberry Pi computers being used everywhere from interactive museum exhibits and schools to national postal sorting offices and government call centres. Kitchen table businesses all over the world have been able to scale and find success in a way that just wasn't possible in a world where integrating technology meant spending large sums on laptops and PCs.

Raspberry Pi removes the high entry cost to computing for people across all demographics: while children can benefit from a computing education that previously wasn't open to them, many adults have also historically been priced out of using computers for enterprise, entertainment, and creativity.

Raspberry Pi eliminates those barriers.

Raspberry Pi Press

store.rpipress.cc

Raspberry Pi Press is your essential bookshelf for computing, gaming, and hands-on making. We are the publishing imprint of Raspberry Pi Ltd. From building a PC to building a cabinet, discover your passion, learn new skills, and make awesome stuff with our extensive range of books and magazines.

The MagPi

magpi.raspberrypi.com

The MagPi is the official Raspberry Pi magazine. Written for the Raspberry Pi community, it is packed with Pi-themed projects, computing and electronics tutorials, how-to guides, and the latest community news and events.

HackSpace

hackspace.raspberrypi.com

HackSpace magazine is filled with projects for fixers and tinkerers of all abilities. We'll teach you new techniques and give you refreshers on familiar ones, from 3D printing, laser cutting, and woodworking to electronics and the Internet of Things. *HackSpace* will inspire you to dream bigger and build better.

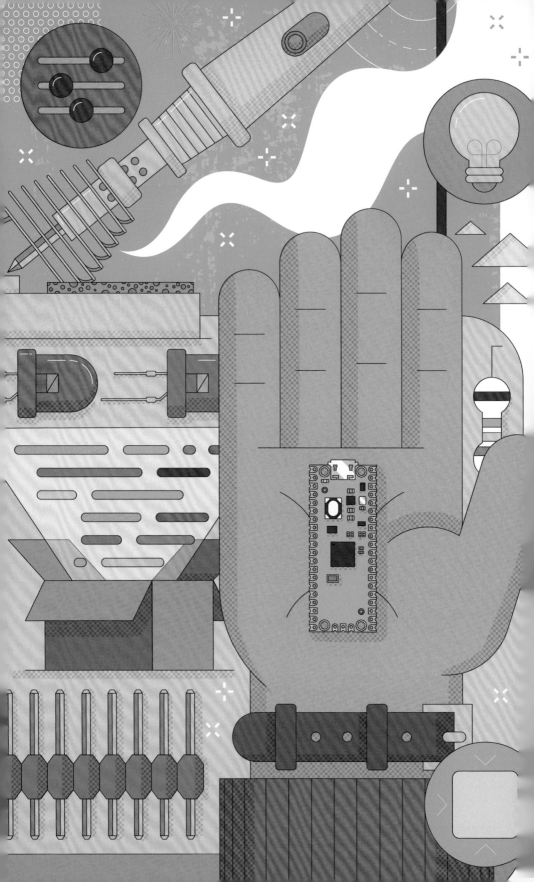

Chapter 1

Get to know your Raspberry Pi Pico

Get acquainted with your powerful new microcontroller board and learn how to attach pin headers and install MicroPython to program it

Raspberry Pi Pico is a miniature marvel, putting the same technology that underpins everything from smart home systems to industrial factories in the palm of your hand. Whether you're looking to learn about the MicroPython programming language, take your first steps in physical computing, or want to build a hardware project, Raspberry Pi Pico — and its amazing community — will support you every step of the way.

Raspberry Pi Pico and Pico W are *microcontroller development boards*. They're designed for experimenting with physical computing using a special type of processor: a *microcontroller*. The size of a stick of gum, Raspberry Pi Pico and Pico W pack a surprising amount of power thanks to the chip at the centre of the board: an RP2040 microcontroller.

Raspberry Pi Pico and Pico W aren't designed to replace Raspberry Pi, which is an entirely different class of device known as a *single-board computer*. You might use Raspberry Pi to play games, write software, or browse the web. Raspberry Pi Pico is designed for physical computing projects, where it is used to control anything from LEDs and buttons to sensors, motors, and even other microcontrollers.

Throughout this book you'll be learning all about Raspberry Pi Pico, but the skills you learn will also apply to any other development board based around its RP2040 microcontroller — and even other devices, so long as they are compatible with the MicroPython programming language.

A guided tour of Raspberry Pi Pico

Raspberry Pi Pico — 'Pico' for short — is a lot smaller than even Raspberry Pi Zero, the most compact of Raspberry Pi's single-board computer family. Despite this, it includes a lot of features — all accessible using the *pins* around the edge of the board. It's available in two versions, Raspberry Pi Pico and Raspberry Pi Pico W; you'll see the difference between the two later.

Figure 1-1 shows Raspberry Pi Pico as seen from above. If you look at the longer edges, you'll see gold-coloured sections with small holes. These are the pins which provide the RP2040 microcontroller with connections to the outside world — known as *input/output* (IO).

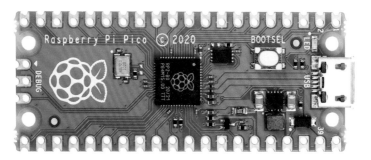

Figure 1-1   The top of the board

The pins on your Pico are very similar to the pins that make up the general-purpose input/output (GPIO) header on a Raspberry Pi — but while most Raspberry Pi single-board computers come with the physical metal pins already attached, Raspberry Pi Pico and Pico W do not.

If you want to buy a Pico with headers mounted, look for Raspberry Pi Pico H and Pico WH instead. There's a good reason to offer models without headers attached: look at the outer edge of the circuit board and you'll see it's bumpy, with little circular cut-outs (**Figure 1-2**).

These bumps create what is called a *castellated circuit board*, which can be soldered on top of other circuit boards without using any physical metal pins. It's very helpful in builds where you need to keep the height to a minimum, making for a smaller finished project. If you buy an off-the-shelf gadget powered by Raspberry Pi Pico or Pico W, it'll almost certainly be fitted using the castellations.

The holes just inwards from the bumps are to accommodate *2.54mm male pin headers*. You'll recognise them as the same type of pins used on the bigger Raspberry Pi's GPIO header. By soldering these in place pointing downwards, you can push your Pico into a *solderless breadboard* to make connecting and disconnecting new hardware as easy as possible — great for experiments and rapid prototyping!

The chip at the centre of your Pico (**Figure 1-3**) is an RP2040 microcontroller. This is a custom *integrated circuit* (*IC*), designed and built by Raspberry Pi to operate as the brains of your Pico and other microcontroller-based devices. If you look at it closely, you'll see a Raspberry Pi logo etched into the top of the chip along with a series of letters and numbers which let engineers track when and where the chip was made.

Figure 1-2
Castellation

Figure 1-3
RP2040 chip

At the top of your Pico is a *micro USB port* (**Figure 1-4**). This provides power to make your Pico run, and also sends and receives data that lets your Pico talk to a Raspberry Pi or another computer via its USB port. This is how you'll load programs onto your Pico.

If you hold your Pico up and look at the micro USB port head-on, you'll see it's shaped so it's narrower at the bottom and wider at the top. Take a micro USB cable, and you'll see its connector is the same.

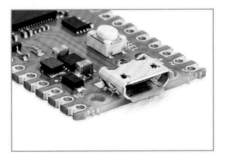

Figure 1-4
micro USB port

The micro USB cable will only go into the micro USB port on your Pico one way up. When you're connecting it, make sure to line the narrow and wide sides up the right way around — you could damage your Pico if you try to brute-force the micro USB cable in the wrong way up!

Just below the micro USB port is a small button marked 'BOOTSEL', shown in **Figure 1-5**. 'BOOTSEL' is short for *boot selection*, which switches your Pico between two start-up modes when it's first switched on. You'll use the boot selection button later, as you get your Pico ready for programming.

At the bottom of your Pico are three smaller gold pads with the word 'DEBUG' above them (**Figure 1-6**). These are designed for debugging, or finding errors, in programs running on the Pico, using a special tool called a *debugger*. You won't need to use the debug header at first, but you may find it useful as you write larger and more complicated programs. On some Raspberry Pi Pico models, the debug pads are replaced by a small, three-pin connector.

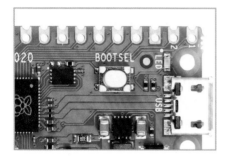

Figure 1-5
Boot selection button

Figure 1-6
Debug pads

Turn your Pico over and you'll see the underside has writing on it (**Figure 1-7**). This printed text is known as a *silk-screen layer*, and labels each of the pins with its core function. You'll see things like 'GP0' and 'GP1', 'GND', 'RUN', and '3V3'. If you ever forget which pin is which, these labels will tell you — but you won't be able to see them when the Pico is pushed into a breadboard, so we've printed full pinout diagrams in this book for easier reference.

You might have noticed that not all the labels line up with their pins. The small holes at the top and bottom of the board are *mounting holes*, designed to allow you to fix your Pico to projects more permanently, using screws or nuts and bolts. Where the holes get in the way of the labelling, the labels are pushed further up or down the board: looking at the top-right. So 'VBUS' is the first pin on the left, 'VSYS' the second, and 'GND' the third.

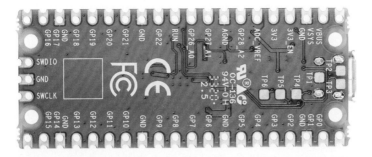

Figure 1-7   Labelled underside

You'll also see some flat, gold pads labelled with 'TP' and a number. These are test points, and are designed for engineers to quickly check that a Raspberry Pi Pico is working after it has been assembled at the factory — you won't be using them yourself. Depending on the test pad, the engineer might use a multimeter or an oscilloscope to check that your Pico is working properly before it's packaged up and shipped to you.

If you have a Raspberry Pi Pico W or Pico WH, you'll find another piece of hardware on the board: a silver metal rectangle (**Figure 1-8**). This is a shield for a wireless module, like the one on Raspberry Pi 4 and Raspberry Pi 5, which can be used to connect your Pico to a Wi-Fi network or to Bluetooth devices. It's connected to a small antenna which sits at the very bottom of the board — which is why you'll find the debug pads or connector closer to the middle of the board on Raspberry Pi Pico W and Pico WH.

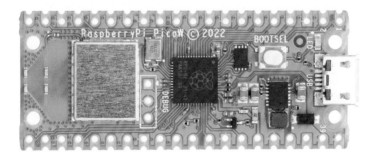

Figure 1-8   The Raspberry Pi Pico W wireless module and antenna

Soldering the headers

When you unpack your Raspberry Pi Pico or Pico W, you'll notice that it is completely flat. There are no metal pins sticking out from the sides, like you'd find on the GPIO header of your Raspberry Pi or on the Pico H and Pico WH. You can use the castellations to attach your Pico to another circuit board, or to solder wires for a project where your Pico will be permanently fixed in place.

The easiest way to use your Pico, though, is to attach it to a *solderless breadboard* — and for that, you'll need to attach *pin headers*. You'll need a soldering iron with a stand, some solder, a cleaning sponge, your Pico, and two 20-pin 2.54 mm male header strips. If you already have a solderless breadboard, you can use it to make the soldering process easier.

Sometimes 2.54 mm headers are provided in strips longer than 20 pins. If yours are longer, just count 20 pins in from one end and look at the plastic between the 20th and 21st pins: you'll see it has a small indentation at either side. This is a *break point*: if you have flush cutters, you can snip them easily. If not, put your thumbnails in the indentation with the headers in both your left and right hands and bend the strip. It will break cleanly, leaving you with a strip of exactly 20 pins. If the remaining header strip is longer than 20 pins, do the same again so you have two 20-pin strips.

WARNING

Soldering irons are not toys: they get very, very hot, and stay hot for a long time after they're unplugged. If you're a younger learner, make sure you have adult supervision; whether you're young or old, make sure that you put the iron in the stand when you're not using it and never ever touch the metal parts — even after it's unplugged.

Turn your Pico upside-down, so you can see the silk-screen pin numbers and test points on the bottom. Take one of the two header strips and push it gently into the pin holes on the left-hand side of your Pico. Make sure that it's properly inserted in the holes, and not just resting in the castellations, and that all 20 pins are in place, then take the other header and insert it into the right-hand side. When you've finished, the plastic blocks on the pins should be pushed up against your Pico's circuit board.

Pinch your Pico at the sides to hold both the circuit board and the two header strips. Don't let go, or the headers will fall out! If you don't have a breadboard yet, you'll need a way to hold the headers in place while you're soldering — and don't use your fingers, or you'll burn them. You can hold the headers in place with small alligator clips, or a small blob of Blu Tack or other sticky putty (**Figure 1-9**). Solder one pin, then check the alignment: if the pins are at an angle, melt the solder as you carefully adjust them to get everything lined up.

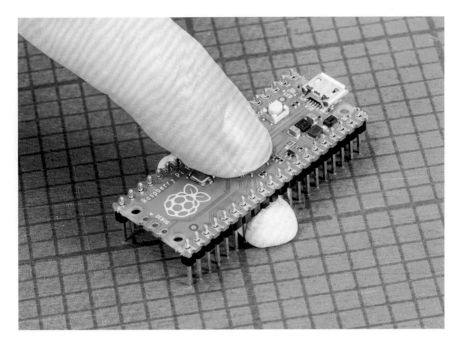

Figure 1-9 You can hold the headers in place with sticky putty before soldering

If you do have a breadboard, simply turn your breadboard and Pico upside down — remembering to keep the headers pinched — and use your Pico to gently push the headers into the holes on your breadboard, taking care to make sure the headers aren't going in at an angle. Keep pushing until your Pico is lying flat, with the plastic blocks on the pin headers sandwiched between your Pico and your breadboard (**Figure 1-10**).

Look at the top of your Pico: you'll see a small length of each pin is sticking up out of the pin holes. This is the part you're going to solder — which means heating up both the pins and the pads on the Pico and melting a small amount of a special metal (solder) onto them.

> **WARNING**
>
> Soldering is a great skill to learn, but it does take practice. Read the directions that follow carefully and in full before even turning your soldering iron on, and remember to take things slowly and carefully. Avoid using too much solder, too: it's easy to add more to a joint with too little solder, but can be harder to take excess solder away — especially if it's splashed over to other parts of your Pico.

Figure 1-10   Alternatively, use a breadboard to hold the headers in place for soldering

Put your soldering iron in its stand, making sure the metal tip isn't resting up against anything, and plug it in. It will take a few minutes for the tip of the iron to get hot; while you're waiting, unroll a small length of solder — about twice as long as your index finger. You should be able to break the solder by pulling and twisting it; it's a very soft metal.

WARNING

While modern solder is widely available in a lead-free formulation, it's still poisonous thanks to a special substance called flux. This is a material which serves to clean the joint and facilitate bonding as you're soldering. It won't harm you if you get it on your fingers, but it could make you ill if you were to eat it, and you should work in a well-ventilated area because it isn't great to inhale, either. Only handle the solder when you're actively using it, and always wash your hands afterwards — especially before you eat anything.

If your soldering stand has a cleaning sponge, take the sponge to the sink and put a little bit of cold water (preferably distilled or deionized) on it so it softens. Squeeze the excess water out of the sponge, so it's damp but not dripping, and put it back on the stand. If you're using a cleaner made of coiled brass wire, you don't need any water.

Pick up your soldering iron by the handle, making sure to keep the cable from catching on anything as you move it around. Hold it like a pencil, but make sure your fingers only ever touch the plastic or rubber handle area: the metal parts, even the shaft ahead of the actual iron tip, will be extremely hot and can burn you very quickly.

Before you begin soldering, clean the iron's tip: brush it along your sponge or coiled wire cleaner. Take your length of solder, holding it at one end, and push the other end onto the tip of your iron: it should quickly melt into a blob. If it doesn't, leave your soldering iron to heat up for longer — or try giving the tip another clean.

Putting a blob of solder on the tip is known as *tinning* the iron. The flux in the solder helps to burn off any oxidation still on the tip of the iron, and gets it ready. Wipe the iron on your sponge or cleaning wire again to clean off the excess solder; the tip should be left looking shiny and clean.

Put the iron back in the stand, where it should always be unless you're actively using it, and move your Pico so it's in front of you. Pick up the iron in one hand and the solder in the other. Press the tip of the iron against the pin closest to you, so that it's touching both the vertical metal pin and the gold-coloured pad on your Pico at the same time (**Figure 1-11**).

Figure 1-11   Heat the pin and pad

It's important that the pin and the pad are both heated up, so keep your iron pressed against both while you count to three. When you've reached three, still keeping the iron in place, press the end of your length of solder gently against both the pin and pad but on the opposite side to your iron tip, as shown in **Figure 1-12**. Just like when you tinned the tip, the solder should melt quickly and begin to flow.

Figure 1-12   Add a little solder

The solder will flow around the pin and the pad, but no further: that's because your Pico's circuit board is coated in a layer called *solder resist* which keeps the solder where it needs to be. Make sure not to use too much solder: a little goes a long way.

If you're using Blu Tack or some other putty, solder the corner pins first to anchor the headers, then remove the putty before you solder any more. That way, you don't have to worry about cooking the putty as you solder.

Pull the remaining part of your solder away from the joint, making sure to keep the iron in place. If you pull the iron away first, the solder will harden and you won't be able to remove the piece in your hand; if that happens, just put the iron back in place to melt it again. Once the molten solder has spread around the pin and pad (**Figure 1-13**), which should only take a second or so, remove the soldering iron. Congratulations: you've soldered your first pin!

Figure 1-13   Now remove the iron

Clean the tip of your iron on your sponge or brass wire, and put it back in the stand. Pick up your Pico and look at your solder joint: it should fill the pad and rise up to meet the pin smoothly, looking a little like a volcano shape with the pin filling in the hole where the lava would be, as shown in **Figure 1-14**.

Figure 1-14   Well-soldered pins

If the solder is too hot, it won't flow well and you'll get an overheated joint with some burnt flux (example **A** in **Figure 1-15**). This can be removed with a bit of careful scraping with the tip of a knife, or a toothbrush and a little 90% isopropyl alcohol.

On the other hand, if the solder is entirely covering the pin, as in example **B** in **Figure 1-15**, you used too much. That's not necessarily going to cause a problem, though it doesn't look very attractive: so long as none of the solder is touching any of the pins around it, it should still work. If it is touching other pins (as in example **C** of **Figure 1-15**), you've created a *bridge* which will cause a short circuit.

Again, bridges are easy to fix. First, try reflowing the solder on the joint you were making; if that doesn't work, put your iron against the pin and pad at the other side of the bridge to flow some of it into the joint there. If there's still far too much solder, you'll need to remove the excess before you use your Pico: you can buy desoldering braid, which you press against the molten solder to wick the excess up, or a desoldering pump to physically suck the molten solder up.

If the solder is sticking to the pin but not sticking to the copper pad, as in example **D** in **Figure 1-15**, then the pad wasn't heated up enough. Don't worry, it's easily fixed: take your soldering iron and place it where the pad and pin meet, making sure that it's pressing against both this time. After a few seconds, the solder should reflow and make a good joint.

Another common mistake is too little solder: if you can still see copper pad, or there's a gap between the pin and the pad which isn't filled in with solder, you used too little (example **E** in **Figure 1-15**). Put the iron back on the pin and pad, count to three, and add a little more solder. Too little is always easier to fix than too much, so remember to take it easy with the solder!

Once you're happy with the first pin, repeat the process for all 40 pins on your Pico — leaving the three-pin 'DEBUG' header at the bottom empty. Tip: solder the four corner pins first. Take your time, don't rush, and remember that mistakes can always be fixed. Remember to clean your iron's tip regularly during your soldering, and if you find things are getting difficult, melt some solder on it to re-tin the tip. Be sure to keep refreshing your length of solder, too: if it's too short and your fingers are too close to the iron's tip, you can burn yourself.

When you're finished, and you've checked all the pins for good solder joints and to make sure they're not bridged to any nearby pins, clean and tin the iron's tip one last time before putting it back in the stand and unplugging it. Make sure to let the iron cool before you put it away: soldering irons can stay hot enough to burn you for a long time after they've been unplugged!

Figure 1-15   Examples of soldering issues

Finally, it's time to wash your hands — and celebrate your new skill as a soldering supremo!

Installing MicroPython

Now that your Pico is ready to go (**Figure 1-16**), there's only one thing left to do to get it ready: install MicroPython onto it. Start by plugging a micro USB cable into the micro USB port on your Pico — make sure it's the right way up before gently pushing it in the rest of the way.

> **NOTE**
>
> To install MicroPython onto your Pico, you'll need to download it from the internet onto a computer running Windows, macOS, or Linux (including a Raspberry Pi) and connect your Pico to the computer in order to finish setting it up. You'll only have to do this once: after MicroPython is installed, it will stay on your Pico unless you decide to replace it with something else in the future.

Figure 1-16   All the pins correctly soldered

Hold down the 'BOOTSEL' button on the top of your Pico. Then, while still holding it down, connect the other end of the micro USB cable to one of the USB ports on your computer. Count to three, then let go of the button.

NOTE

On macOS, you may be asked whether you want to "Allow accessory to connect" when you plug the Pico into your computer. You will need to click **Allow** to permit it. After you install MicroPython onto your Pico, macOS may ask the question a second time, because it now looks like a different device.

After a few more seconds you should see your Pico appear as a removable drive, as though you'd connected a USB flash drive or external hard drive.

On a Raspberry Pi, you'll see a pop-up asking if you'd like to open the drive in the File Manager. Make sure **Open in File Manager** is selected and click **OK**.

On Windows, you may see an autoplay notification. You can click that and them choose **Open Folder to View Files**. Alternatively, you can open File Explorer, navigate to **This PC**, and double-click the **RPI-RP2** drive to open it.

On a Mac, it's likely to quietly mount the drive without fanfare. Open the Finder and look for **RPI-RP2** in the sidebar to the left of the Finder window. It's likely to appear under Locations. If the sidebar is not visible, click **View** and select **Show Sidebar**.

In the File Manager window, you'll see two files on your Pico (**Figure 1-17**): **INDEX.HTM** and **INFO_UF2.TXT**. The second file contains information about your Pico, such as the version of the bootloader it's currently running. The first file, **INDEX.HTM**, is a link to the Raspberry Pi Pico website. Double-click on this file or open your web browser and type **rptl.io/microcontroller-docs** into the address bar.

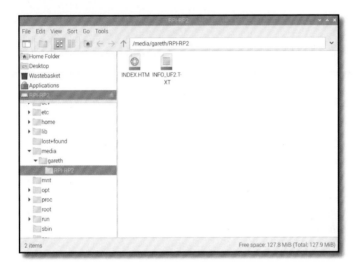

Figure 1-17   You'll see two files on your Raspberry Pi Pico

When the web page opens, you'll see information about Raspberry Pi's microcontroller and development boards, including Raspberry Pi Pico and Pico W. Click on the MicroPython box to go to the firmware download page. Scroll down to the section labelled **Drag-and-Drop MicroPython**, as shown in **Figure 1-18**, and find the link for the version of MicroPython for your board. There's one for Raspberry Pi Pico and Pico H, and another for Raspberry Pi Pico W and Pico WH. Click on the link to download the appropriate UF2 file. If you accidentally download the wrong file, don't worry; you can come back to the page at any time and flash new firmware onto your device using the same process.

Open a new File Manager (Raspberry Pi), Windows Explorer, or macOS Finder window, then navigate to your **Downloads** folder and find the file you just downloaded. it will be called **rp2-pico** or **rp2-pico-w** followed by a date, some identifying text and numbers, along with the extension **uf2**.

Figure 1-18   Click the link to download the MicroPython firmware

Click and hold the mouse button on the UF2 file, then drag it to the other window that's open on your Pico's removable storage drive. Hover it over that window and let go of the mouse button to drop the file onto your Pico, as shown in **Figure 1-19**.

After a few seconds you'll see your Pico drive window disappear from File Manager, Explorer, or Finder, and you may also see a warning that a drive was removed without being ejected. Don't worry, that's supposed to happen! When you dragged the MicroPython firmware file onto your Pico, you told it to flash the firmware onto its internal storage. To do that, your Pico switches out of the special mode you put it in with the 'BOOTSEL' button, flashes the new firmware, and then loads it — your Pico is now running MicroPython.

Congratulations: you're now ready to get started with MicroPython on your Raspberry Pi Pico!

Figure 1-19 Drag the MicroPython firmware file to your Raspberry Pi Pico

FURTHER READING

The webpage linked from INDEX.HTM isn't just a place to download MicroPython. It also hosts plenty of additional resources. Click on the tabs and scroll to access guides, projects, and the databook collection — a bookshelf of detailed technical documentation covering everything from the inner workings of the RP2040 microcontroller which powers your Pico, to programming it in both the Python and C/C++ languages.

Chapter 2

Programming with MicroPython

Connect a computer and start writing programs for your Raspberry Pi Pico using the MicroPython language

Since its launch in 1991, the Python programming language — named after the famous comedy troupe Monty Python, rather than the snake — has grown to become one of the most popular in the world. Its popularity, though, doesn't mean there aren't improvements that could be made — particularly if you're working with a microcontroller.

The Python programming language was developed for computer systems like desktops, laptops, and servers. Microcontroller boards like Raspberry Pi Pico are smaller, simpler, and with considerably less memory — meaning they can't run the same Python language as their bigger counterparts.

That's where MicroPython comes in. Originally developed by Damien George and first released in 2014, MicroPython is a Python-compatible programming language developed specifically for microcontrollers. It includes many of the features of mainstream Python, while adding a range of new ones designed to take advantage of the facilities available on Raspberry Pi Pico and other microcontroller boards.

If you've programmed with Python before, you'll find MicroPython immediately familiar. If not, don't worry: it's a friendly language to learn!

Introducing the Thonny Python IDE

Before you can start to program Pico with MicroPython, you'll need to set up what is called an *integrated development environment (IDE)*. Thonny, a popular IDE for Python and MicroPython, comes preloaded on Raspberry Pi OS and is available for Linux, Windows, and Mac.

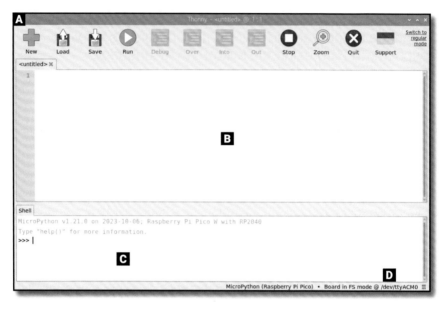

A Toolbar   **C** Python shell

B Script area   **D** Interpreter

The toolbar (**A**) offers an icon-based quick-access system to commonly used program functions — like saving, loading, and running programs. The script area (**B**) is where your Python programs are written. It is split into a main area for your program and a small side margin for showing line numbers.

The Python Shell (**C**) allows you to type individual instructions which are run as soon as you press the **ENTER** key, and also provides information about running programs. This is also known as a *REPL*, for *read-evaluate-print loop*.

The bottom-right of the Thonny window (**D**) shows, and lets you change, the current Python *interpreter* — the version of Python used to run your programs.

THONNY MODES

The view shown is Thonny's Simple mode, which is the default mode on the version of Thonny that comes preinstalled on Raspberry Pi. If you install Thonny on another operating system, the installer will default to Regular mode unless you change it. Regular mode has a more compact toolbar and a full menu.

You can switch from Simple to Regular mode by clicking **Switch to regular mode**, and you can switch from Regular to Simple mode by choosing **Tools → Options → General** and changing the UI mode.

Connecting Thonny to Pico

If you're using Pico with a Raspberry Pi, Thonny is already installed; if you're using a different Linux distribution, Windows, or macOS, open your web browser, visit **thonny.org**, and click the download link at the top of the page to download the Thonny and Python bundle installer for your operating system.

As an integrated development environment, Thonny gathers together, or *integrates*, all the different tools you need to write, or *develop*, software into a single user interface, or *environment*. There are many different IDEs: some allow you to develop in multiple different programming languages while others, like Thonny, focus on a single language.

If you haven't already done so, take your Pico and connect a micro USB cable between it and one of your computer's USB ports — it doesn't matter which one.

Begin by loading Thonny: on Raspberry Pi OS, you can load it by clicking on the Raspberry Pi menu at the top-left on your screen, moving the mouse to the **Programming** section, and clicking on **Thonny**. On Windows, Thonny will be available from the Start menu after you complete installation. On macOS, you can find it in your Applications folder or run it from Launchpad.

PYTHON PROFESSIONALS

If you've worked through the Python chapter in The Official Raspberry Pi Beginner's Guide, much of what you'll read in this chapter will be very familiar. Still, work through the first couple of examples, to get used to the differences in running programs in the Python interpreter on your Raspberry Pi and in the MicroPython interpreter on your Pico, then feel free to skip to the next chapter.

With your Pico connected to your Raspberry Pi, click on the words **Local Python 3** at the bottom-right of the Thonny window. This shows your current interpreter, which is responsible for taking the instructions you type and turning them into code that the computer, or microcontroller, can understand and run. Normally the interpreter is the copy of Python running on your Raspberry Pi, but it needs to be changed to run your MicroPython programs on your Pico.

Look for **MicroPython (Raspberry Pi Pico)** (**Figure 2-1**) in the list that appears, and click on it. If you can't see it in the list, double-check that your Pico is properly plugged into the micro USB cable, and that the micro USB cable is properly plugged into your Raspberry Pi or other computer.

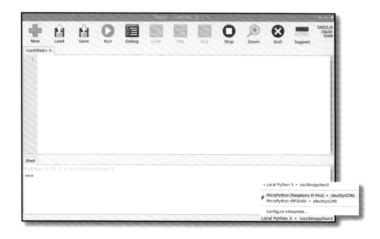

Figure 2-1   Choosing a Python interpreter

Look at the Python Shell at the bottom of the Thonny window: you'll see that it now reads **MicroPython** and tells you that it's running on **Raspberry Pi Pico**. Congratulations: you're ready to start programming.

> **? INTERPRETER SWITCHING**
>
> Choosing the interpreter picks where and how your program will run: when you choose **MicroPython (Raspberry Pi Pico)**, programs will run on your Pico; picking **Local Python 3** means programs will run on your Raspberry Pi or computer instead.
>
> If you find programs aren't running where you'd expect, make sure to check which interpreter Thonny is set to use!

Your first MicroPython program: Hello, World!

To start writing your first program, click on the Python shell area at the bottom of the Thonny window, just to the right of the bottom **>>>** symbols, and type the following instruction before pressing the **ENTER** key:

```
print("Hello, World!")
```

When you press **ENTER**, you'll see that your program begins to run instantly: Python will respond, in the same shell area, with the message 'Hello, World!' (**Figure 2-2**), just as you asked. That's because the shell is a direct line to the MicroPython interpreter running on your Pico, whose job it is to look at your instructions and interpret what they mean. This interactive mode works the same as when you're programming your Raspberry Pi: instructions written in the shell area are acted on immediately, with no delay. The only difference: they're sent to your Pico to run them, and any result — in this case the message 'Hello, World!' — is sent back to your Raspberry Pi or computer to be displayed.

Figure 2-2   MicroPython prints the 'Hello, World!' message in the shell area

---

SYNTAX ERROR

If your program doesn't run but instead prints a 'syntax error' message to the shell area, there's a mistake somewhere in what you've written. Python needs its instructions to be written in a very specific way: if you miss a bracket or a quotation mark, spell 'print' wrong or give it a capital P, or add extra symbols somewhere in the instruction, your program won't run. Try typing the instruction again, and make sure it matches the version in this book before pressing the **ENTER** key.

Programming your Pico using the Shell is a little like having a telephone conversation: when you press the **ENTER** key, your instruction is sent through the micro USB cable to the MicroPython interpreter running on your Pico; the interpreter looks at your instruction, does whatever it is told, then sends the result back through the micro USB cable to Thonny.

You don't have to program your Pico (or your Raspberry Pi) in interactive mode. Click on the script area in the middle of the Thonny window, then type your program again:

```
print("Hello, World!")
```

When you press the **ENTER** key this time, nothing happens — except that you get a new, blank line in the script area. To make this version of your program work, you'll have to click the **Run** icon ⊙ in the Thonny toolbar.

Even though this is a simple program, you'll want to get in the habit of saving your work. Before you run your program, click the **Save** icon 💾. You'll be asked whether you want to save your program to **This computer**, meaning your Raspberry Pi or whatever other computer you're running Thonny on, or to **Raspberry Pi Pico** (Figure 2-3). Click **Raspberry Pi Pico**, then type a descriptive name like **Hello World.py** and click the OK button.

Figure 2-3   Saving a program to Pico

Click the **Run** icon ⊙ now. It will run automatically on your Pico. You'll see two messages appear in the shell area at the bottom of the Thonny window:

```
>>> %Run 'Hello World.py'
 Hello, World!
```

The first of these lines is an instruction from Thonny telling the MicroPython interpreter on your Pico to run the code that's in the file you saved. The second is the output of the program — the message you told MicroPython to print. Congratulations: now you've written two MicroPython programs, in interactive and script modes, and you've successfully run them on your Pico!

There's just one more piece to the puzzle: loading your program again. Close Thonny by pressing the X at the top-right of the window on Windows or Linux (use the close button at the top-left of the window on macOS), then launch Thonny again. This time, instead of writing a new program, click the **Load** icon 📂 in the Thonny toolbar. You'll be asked whether you want to load from **This computer** or your **Raspberry Pi Pico** again. Click **Raspberry Pi Pico** and you'll see a list of all the programs you've saved to your Pico.

When you tell Thonny to save your program on the Pico, it means that the programs are stored on the Pico itself. If you unplug your Pico and plug it into a different computer, your programs will still be where you saved them: on your very own Pico!

Find **Hello World.py** in the list — if your Pico is new, it will be the only file there. Click to select it, then click OK. Your program will load into Thonny, ready to be edited, or for you to run it again.

Next steps: loops and code indentation

A MicroPython program, just as with a standard Python program, normally runs top-to-bottom: it goes through each line in turn, running it through the interpreter before moving on to the next, just as if you were typing them line-by-line into the Shell.

A program that just runs through a list of instructions line-by-line wouldn't be very clever, though — so MicroPython, just like Python, has its own way of controlling the sequence in which its programs run: *indentation*.

Create a new program by clicking on the **New** icon in the Thonny toolbar. You won't lose your existing program; instead, Thonny will create a new tab above the script area. Start your program by typing in the following two lines:

```
print("Loop starting!")
for i in range(10):
```

The first line prints a simple message to the Shell, just like your Hello World program. The second begins a *definite* loop, which will repeat (*loop*) one or more instructions a set number of times. A *variable*, **i**, is assigned to the loop and given a series of numbers to count — using the **range** instruction, which is told to start at the number 0 and work upwards towards, but never reaching, the number 10. The colon symbol (**:**) tells MicroPython that the *body* of the loop begins on the next line.

Variables are powerful tools: as their name suggests, variables are values which can change — or vary — over time and under the control of the program. At its most simple, a variable has two aspects: its name, and the *data* it stores. In the case of your loop, the variable's name is **i** and its data is set by the **range** instruction — starting at 0 and increasing by 1 each time the loop finishes and begins afresh.

To include a line of code in the body of the loop, it has to be *indented* — moved in from the left-hand side of the script area. The next line starts with four blank spaces, which Thonny will have added automatically when you pressed **ENTER** after line 2. Type it in now:

```
    print("Loop number", i)
```

The four blank spaces push this line inwards compared to the other lines in your program. This indentation is how MicroPython tells the difference between instructions outside the loop and instructions inside the loop: the indented code, forming the inside of the loop, is known as being *nested*.

You'll notice that when you pressed **ENTER** at the end of the third line, Thonny automatically indented the next line — assuming it would be part of the loop. To remove this indentation, just press the **BACKSPACE** key once before typing the fourth line:

```python
print("Loop finished!")
```

Your four-line program is now complete. The first line sits outside the loop, and will only run once; the second line sets up the loop; the third sits inside the loop and will run once for each time the loop loops; and the fourth line sits outside the loop once again.

```python
print("Loop starting!")
for i in range(10):
    print("Loop number", i)
print("Loop finished!")
```

Click the **Save** icon and choose to save the program on your Pico and call it **Indentation.py**. Next, click the **Run** icon. The program will run as soon as it is saved: look at the Shell area for its output (**Figure 2-4**).

```
Loop starting!
Loop number 0
Loop number 1
Loop number 2
Loop number 3
Loop number 4
Loop number 5
Loop number 6
Loop number 7
Loop number 8
Loop number 9
Loop finished!
```

COUNT FROM ZERO

Python is a zero-indexed language — meaning it starts counting from 0, not from 1. This is why your program prints the numbers starting at 0 rather than 1. Also, the upper limit you pass to the range instruction is exclusive, which means that **range** stops counting when it reaches 9 rather than 10.

If you wanted to count from 1 to 10, you could change this behaviour by switching the **range(10)** instruction to include a starting number, and specify 11 as the upper limit: **range(1, 11)** — or any other numbers you like. If you don't supply a starting number, Python will count from 0 to the integer that's one less than the upper limit.

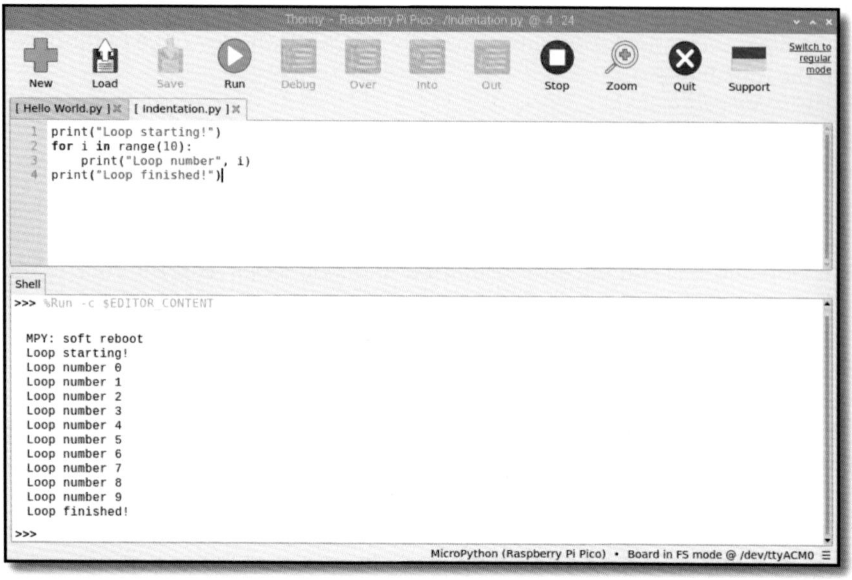

Figure 2-4   Executing a loop

Indentation is one of the most common reasons for a program to not work as you expected. When looking for problems in a program, a process known as *debugging*, always double-check the indentation — especially when you begin nesting loops within loops.

MicroPython also supports *infinite* loops, which run without end. To change your program from a definite loop to an infinite loop, edit line 2 to read:

```
while True:
```

Since we'll no longer be using the variable **i**, change line 3 to read:

```
print("Loop running!")
```

To avoid the program running too quickly, we'll also add a short time delay by importing the *time* library at the start and adding a one-second sleep delay to the loop (you'll learn more about this library in later chapters). Your program should now look like this:

```
import time
print("Loop starting!")
while True:
    print("Loop running!")
```

```
    time.sleep(1)
print("Loop finished!")
```

Click the **Run** icon again, and you'll see the 'Loop starting!' message followed by a never-ending string of 'Loop running!' messages (**Figure 2-5**). The 'Loop finished!' message will never print, because the loop has no end: every time Python has finished printing the 'Loop running!' message, it goes back to the beginning of the loop and prints it again.

Click the **Stop** icon on the Thonny toolbar to tell the program to stop what it's doing — known as *interrupting* the program — and to restart the MicroPython interpreter. You'll see a message appear in the Shell area and the program will stop, without ever reaching line 6.

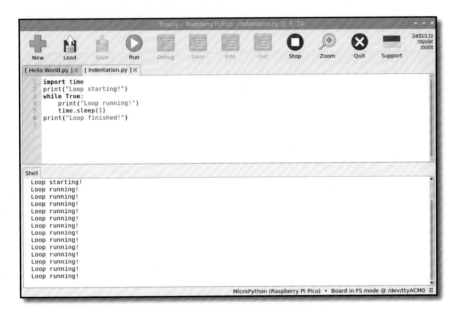

Figure 2-5   An infinite loop, which keeps going until you stop the program

---

CHALLENGE: LOOP THE LOOP

Can you change the loop back into a definite loop again? Can you add a second definite loop to the program? How would you add a loop within a loop, and how would you expect that to work?

Conditionals and variables

Variables in MicroPython, as in all programming languages, exist for more than just controlling loops. Start a new program by clicking the **New** icon on the Thonny toolbar, then type the following into the script area:

```
user_name = input("What is your name? ")
```

Click the **Save** icon, choose to save the program on your Pico and call it **Name Test.py**. Next, click the **Run** icon and watch what happens in the Shell area: you'll be asked for your name. Click into the Shell area to give it focus. Next, type your name into the Shell area, followed by **ENTER**. Because that's the only instruction in your program, nothing else will happen (**Figure 2-6**). If you want to actually do anything with the data you've placed into the variable, you'll need more lines in your program.

Figure 2-6   The input function lets you ask a user for some text input

USING = AND ==

The key to using variables is to learn the difference between = and ==. Remember: = means 'make this variable equal to this value', while == means 'check to see if the variable is equal to this value'. Mixing them up is a sure way to end up with a program that doesn't work!

To make your program do something useful with the name, add a *conditional* statement by typing the following from line 2 onwards:

```
if user_name == "Clark Kent":
    print("You are Superman!")
else:
    print("You are not Superman!")
```

Remember that when Thonny sees that your code needs to be indented, it will do so automatically — but it doesn't know when your code needs to stop being indented, so you'll have to delete the spaces yourself.

Click the **Run** icon and type your name into the Shell area. Unless your name happens to be Clark Kent, you'll see the message 'You are not Superman!'. Click **Run** again, and this time type in the name 'Clark Kent' — making sure to write it exactly as in the program, with a capital C and K. This time, the program recognises that you are, in fact, Superman (**Figure 2-7**).

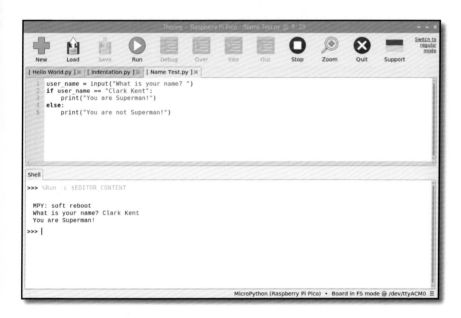

Figure 2-7   Shouldn't you be out saving the world?

The **==** symbols tell Python to do a direct comparison, looking to see if the variable **user_name** matches the text — known as a *string* — in your program. If you're working with numbers, there are other comparisons you can make: **>** to see if a number is greater than another number, **<** to see if it's less than, **>=** to see if it's greater than or equal to, **<=** to see if it's less than or equal to. There's also **!=**, which means not equal to — it's the exact opposite of **==**. These symbols are technically known as *comparison operators*.

Comparison operators can also be used in loops. Delete lines 2 through 5, then type the following in their place:

```
while user_name != "Clark Kent":
    print("You are not Superman - try again!")
    user_name = input("What is your name? ")
print("You are Superman!")
```

CHALLENGE: ADD MORE QUESTIONS

Can you change the program to ask more than one question, storing the answers in multiple variables? Can you make a program which uses conditionals and comparison operators to print whether a number typed in by the user is higher or lower than 5?

Click the **Run** icon again. This time, rather than quitting, the program will keep asking for your name until it confirms that you are Superman (**Figure 2-8**) — sort of like a very simple password. To get out of the loop, either type 'Clark Kent' into the script area or click the **Stop** icon on the Thonny toolbar. Congratulations: you now know how to use conditionals and comparison operators!

Figure 2-8   The program will keep asking for your name until you say it's 'Clark Kent'

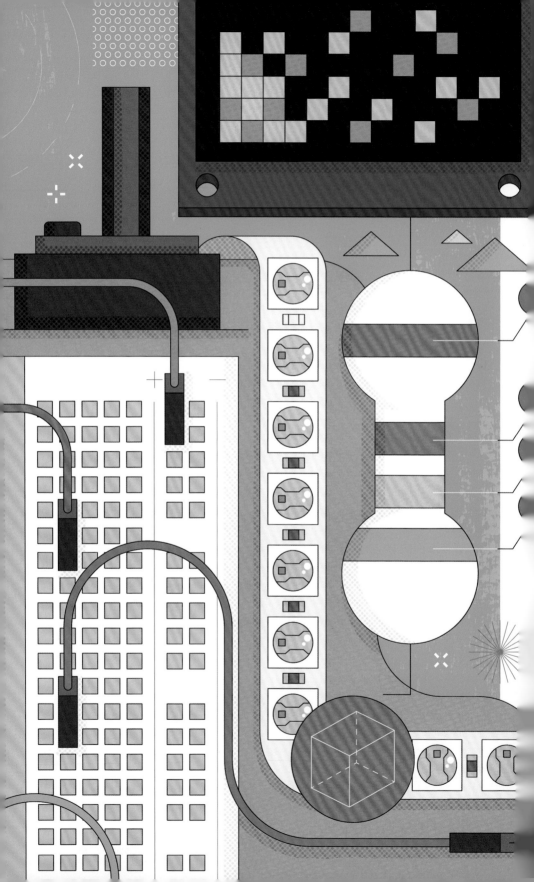

Chapter 3

Physical computing

Learn about your Raspberry Pi Pico's pins and the electronic components you can connect and control

When people think of 'programming' or 'coding', they're usually — and naturally — thinking about software. Coding can be about more than just software, though: it can affect the real world through hardware. This is called *physical computing*. As the name suggests, physical computing is all about controlling things in the real world with your programs: hardware, rather than software. When you set the program on your washing machine, change the temperature on your programmable thermostat, or press a button at traffic lights to cross the road safely, you're using physical computing.

These devices are typically controlled by a microcontroller very much like the one on your Raspberry Pi Pico — and it's entirely possible for you to create your own control systems by learning to take advantage of your Pico's capabilities, just as easily as you learned to write software that runs on your Pico.

Your Pico's pins

Your Pico talks to hardware through the series of pins along both its edges. Most work as programmable input/output (PIO) pins, meaning they can be programmed to act as either an input or an output, and have no preset purpose of their own until you assign one. Some pins have extra features and alternative modes for communicating with more complicated hardware; others have a specific purpose, providing connections for things like power.

Raspberry Pi Pico's 40 pins are labelled on the underside of the board, with three also labelled with their numbers on the top of the board: Pin 1, Pin 2, and Pin 39. These top labels help you remember how the numbering works: Pin 1 is at the top-left as you look at the board from above, with the micro USB port to the upper side. Pin 20 is the bottom-left, Pin 21 the bottom-right, and Pin 39 one below the top-right with the unlabelled Pin 40 above it. The labelling on the underside is more thorough, but you won't be able to see it when your Pico is plugged into a breadboard!

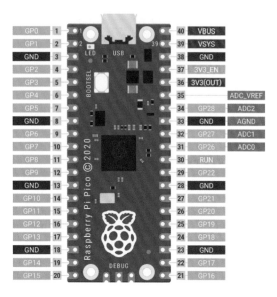

Figure 3-1 The Raspberry Pi Pico's pins, seen from the top of the board

On the Raspberry Pi Pico, pins are usually referred to by their functions (see **Figure 3-1**) rather than by number. There are several categories of pin types, each of which has a particular function:

▸ **3V3(OUT)** — *3.3 volts power* — A source of 3.3V power generated from the VSYS input. This power supply can be switched off by shorting the pin above it (3V3_EN) to GND, which also switches your Pico off.

▸ **VSYS** — *~2-5 volts power* — A pin directly connected to your Pico's internal power supply, which cannot be switched off without also switching the Pico off.

▸ **VBUS** — *5 volts power* — A source of 5V power taken from your Pico's micro USB port, and used to power hardware which needs more than 3.3V. If you are connection the output of a component to your Pico's GPIO pins, take care that the component's output pins do not exceed 3.3V.

- **GND** — *0 volts ground* — A ground connection, used to complete a circuit connected to a power source. Several GND pins are dotted around your Pico to make wiring easier.

- **GPxx** — *General-purpose input/output pin number 'xx'* — The GPIO pins available for your program, labelled GP0 through to GP28.

PIN GP0

Like counting in Python, your Pico's GPIO pins start at the number 0 rather than the number 1. Labelled on the underside of the board, they go from 0 to 28, although some of them aren't broken out as physical pins. Although GP29 is present on the RP2040 microcontroller, it isn't broken out on the Pico board at all, but it is used for an internal board function.

- **GPxx_ADCx** — *General-purpose input/output pin number 'xx', with analogue input number 'x'* — A GPIO pin which ends in ADC and a number can be used as an analogue input as well as a digital input or output — but not both at the same time.

- **ADC_VREF** — *Analogue-to-digital converter (ADC) voltage reference* — A special input pin which sets a reference voltage for any analogue inputs.

- **AGND** — *ADC 0 volts ground* — A special ground connection for use with the ADC_VREF pin.

- **RUN** — *Enables or disables your Pico* — The RUN pin is used to start and stop your Pico from another microcontroller or other controlling device.

Several of the GPIO pins have additional functions, which you'll learn about later in the book. For a full pinout including these additional functions, see Appendix B, *Pinout guide*.

MISSING PINS

The general-purpose input/output pins on Pico are numbered based on the pins of the chip which powers it, an RP2040 microcontroller. Not all the pins available on RP2040 are brought out to your Pico's pins, however — which is why there's a gap in the numbering between the last basic general-purpose pin GP22 and the first analogue-capable pin GP26_ADC0.

Electronic components

Your Pico is only part of what you'll need to work with physical computing. You'll also need some electrical components, the devices you'll control from Pico's GPIO pins. There are thousands of different components available, but most physical computing projects are made using the following common parts.

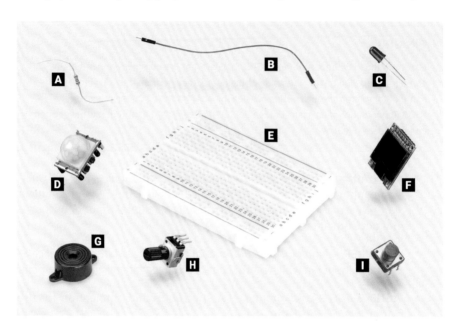

Figure 3-2 Common electronic components

A	Resistor	**F**	OLED display
B	Jumper wire	**G**	Piezoelectric buzzer
C	Light-emitting diode (LED)	**H**	Potentiometer
D	Passive Infrared Sensor (PIR)	**I**	Push-button switch
E	Breadboard		

> ▸ **Resistor (A)** — these are components which control the flow of *electrical current* and are available in different values, measured using a unit called *ohms* (Ω). The higher the number of ohms, the more resistance is provided. For Pico physical computing projects, their most common use is to protect LEDs from drawing too much current and damaging themselves or your Pico; for this you'll want resistors rated at around 330Ω, though many electrical suppliers sell handy packs containing various commonly used values to give you more flexibility.

- **Jumper wires (B)** — also known as *jumper leads*, connect components to your Pico and, if you're not using a breadboard, to each other. They are available in three versions: male-to-female (M2F); female-to-female (F2F), which can be used to connect individual components to your Pico if you're not using a breadboard; and male-to-male (M2M), which is used to make connections from one part of a breadboard to another. Depending on your project, you may need all three types of jumper wire. If you're using a breadboard, you can usually get away with just M2F and M2M jumper wires.

- **Light-emitting diode (LED, C)** — this is an *output device* which you can control directly from your program. An LED lights up when it's powered on, and you'll find them all over your house: from the small ones which let you know when you've left your washing machine switched on, to the large ones you might have lighting up your rooms. LEDs are available in a wide range of shapes, colours, and sizes, but not all are suitable for use with your Pico: avoid any which say they are designed for 5V or 12V power supplies.

- **Passive infrared sensor (PIR, D)** — this is one of a variety of input devices known as *sensors*, designed to report on changes in whatever they are monitoring. In the case of a PIR sensor, it monitors movement of people or animals: the sensor watches for movement in its *field of view* (determined by its plastic lens) and sends a signal when it detects a change. PIR sensors are commonly found on burglar alarms, to find people moving in the dark.

- **Breadboard (E)** — also known as a *solderless breadboard*, can make physical computing projects considerably easier. Rather than having a bunch of separate components which need to be connected with wires, a breadboard lets you insert components and have them connected through metal tracks which are hidden beneath its surface. Many breadboards also include sections for power distribution, making it even easier to build your circuits. You don't need a breadboard to get started with physical computing, but it certainly helps.

- **OLED display (F)** — this is a screen which talks to your Pico over a special communication system such as the *inter-integrated circuit (I2C)* bus. Such a bus lets your Pico control the display panel, sending everything from writing to pictures for it to display. There are lots of types of display available, though a popular one — and the one found in this book — is based around the SSD1306 OLED driver, which supports both I2C and *serial peripheral interface (SPI)* interfaces. Note that some displays only use the I2C bus rather than SPI; they'll still work with your Pico, but will only support the one bus and won't work with the SPI example in this book.

- **Piezoelectric buzzer (G)** — also called a buzzer or a sounder, is another output device. Whereas an LED produces light, a buzzer produces a buzzing noise. Inside the buzzer's plastic housing are a pair of metal plates. When active, these plates vibrate against each other to produce the buzzing sound. There are two types of buzzers: *active buzzers* and *passive buzzers*. Make sure to get an active buzzer, as these are the simplest to use.

- **Potentiometer (H)** — this is the sort of component you might find as a volume control on a music player, and can work as two different components. With two of its three legs connected, it acts as a variable resistor or *varistor*, a type of resistor which can be adjusted at any time by twisting the knob. With all three legs properly wired up, it becomes a *voltage divider* and outputs anything from 0 V to the full voltage input depending on the position of the knob.

- **Push-button switch (I)** — this is the type of switch you might find on controllers for a game console. Commonly available with two or four legs — either type will work with your Pico — the push-button switch is an input device: you can tell your program to wait until you press it and then perform a task. A common variant is a *latching switch*: while a *momentary* push-button is only active when you're holding it down, a latching switch — like a light switch — activates when you toggle it once, then stays active until you toggle it again.

Other common electrical components include motors, which need a special control board before they can be connected to your Pico, infrared sensors which detect movement, temperature and humidity sensors which can be used to predict the weather; and light-dependent resistors (LDRs) — input devices which operate like a reverse LED by detecting light.

Sellers all over the world provide components for physical computing with Raspberry Pi Pico, either as individual parts or in kits which provide everything you need to get started. To find sellers, visit **rptl.io/products**, click **Raspberry Pi Pico series**, and click the **Buy now** button to see a list of Raspberry Pi partner online stores and approved resellers for your country or region.

To complete the projects in this book, you should have at least:

- A Raspberry Pi Pico (or Pico W) with male headers attached

- A micro USB cable

- A solderless breadboard

- A Raspberry Pi or other computer for programming

- Male-to-female (M2F) and male-to-male (M2M) jumper wires

- 3 × single-colour LEDs: red, green, and yellow or amber

- 1 × active piezoelectric buzzer

- 1 × 10 kΩ potentiometer, linear or logarithmic

- 3 × 330 Ω resistors

- At least one HC-SR501 PIR sensor

- 1 × SSD1306 OLED module

- WS2812B RGB LEDs (or compatible)

You will also find it helpful to buy a cheap storage box with multiple compartments, so you can keep the components you're not using in your project safe and tidy. If you can, try to find one that will also fit the breadboard — that way you can tidy everything away each time you're done.

Reading resistor colour codes

Resistors come in a wide range of values, from zero-resistance versions which are effectively just pieces of wire to high-resistance versions the size of your leg. Very few of these resistors have their values printed on them in numbers. Instead, they use a special code (**Figure 3-3**) printed as coloured stripes or bands around the body of the resistor.

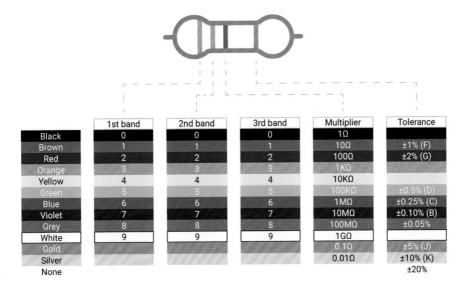

	1st band	2nd band	3rd band	Multiplier	Tolerance
Black	0	0	0	1Ω	
Brown	1	1	1	10Ω	±1% (F)
Red	2	2	2	100Ω	±2% (G)
Orange	3	3	3	1KΩ	
Yellow	4	4	4	10KΩ	
Green	5	5	5	100KΩ	±0.5% (D)
Blue	6	6	6	1MΩ	±0.25% (C)
Violet	7	7	7	10MΩ	±0.10% (B)
Grey	8	8	8	100MΩ	±0.05%
White	9	9	9	1GΩ	
Gold				0.1Ω	±5% (J)
Silver				0.01Ω	±10% (K)
None					±20%

Figure 3-3   Resistor color codes

To read the value of a resistor, position it so the group of bands is to the left and the lone band is to the right. Starting from the first band, look its colour up in the '1st/2nd Band' column of the table to get the first and second digits. This example has two orange bands, which both mean a value of '3' for a total of '33'. If your resistor has four grouped bands instead of three, note down the value of the third band too (for five/six-band resistors, see **rptl.io/5-6-band**).

Moving onto the last grouped band — the third or fourth — look its colour up in the 'Multiplier' column. This tells you what you need to multiply your current number by to get the actual value of the resistor. This example has a brown band, which means '×10^1'. That may look confusing, but it's simply *scientific notation*: '×10^1' simply means 'add one zero to the end of your number'. If it were blue, for ×10^6, you would add six zeroes instead.

Taking 33 from the orange bands, plus the added zero from the brown band, gives us 330 — which is the value of the resistor, measured in ohms. The final band, the one on the right, is the *tolerance* of the resistor. This is simply how close to its rated value it is likely to be. Cheaper resistors might have a silver band, indicating a tolerance 10% higher or lower than its rating, or no last band at all, indicating a tolerance 20% higher or lower. The most expensive resistors have a grey band, indicating a tolerance within 0.05% of its rating. For hobbyist projects, accuracy isn't that important: any tolerance will usually work fine.

If your resistor value goes above 1000 ohms (1000Ω), it is usually rated in kilohms (kΩ); if it goes above a million ohms, those are megohms (MΩ). A 2200Ω resistor would be written as 2.2kΩ; a 2,200,000Ω resistor would be written as 2.2MΩ.

CAN YOU WORK IT OUT?

What colour bands would a 100Ω resistor have? What colour bands would a 2.2MΩ resistor have? If you wanted to find the cheapest resistors, what colour tolerance band would you look for?

Chapter 4

Physical computing with Raspberry Pi Pico

Start connecting basic electronic components to Raspberry Pi Pico and writing programs to control and sense them

Raspberry Pi Pico, with its RP2040 microcontroller, is designed with physical computing in mind. Its numerous general-purpose input/output (GPIO) pins let it talk to a range of components, allowing you to build up projects, from lighting LEDs to recording data about the world around you.

Physical computing is no more difficult to learn than traditional computing: if you could follow the examples in Chapter 2, *Programming with MicroPython*, you'll be able to build your own circuits and program them to interact with the real world.

Your first physical computing program: Hello, LED!

Just as printing 'Hello, World' to the screen is the usual first step in learning a programming language, making an LED light up is the traditional introduction to learning physical computing on a new platform. You can get started without any additional components, too: your Raspberry Pi Pico has a small LED, known as a *surface-mount device (SMD) LED*, on top.

Start by finding the LED: it's the small rectangular component to the left of the micro USB port at the top of the board (**Figure 4-1**), marked 'LED'.

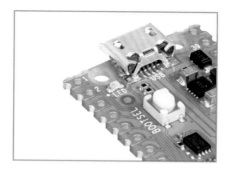

Figure 4-1
The on-board LED is found to the left of the micro
USB connector

The on-board LED is connected to a GPIO pin (GP25 for Pico, but a GPIO on the wireless chip for Pico W) that is not broken out to a physical pin on the edge of your Pico. While you can't connect any hardware to the pin (other than the on-board LED), it can be treated just the same as any other GPIO pin within your programs, but must be referred to as **"LED"**. It's a simple way to add an output to your programs without needing any extra components.

Load Thonny and, if you haven't already done so, configure it to connect to your Pico as shown in "Connecting Thonny to Pico" on page 21. Click the **New** icon to start a new program if needed, then click into the script area, and start your program with the following line:

```
import machine
```

This short line of code is key to working with MicroPython on your Pico. It loads, or *imports*, a collection of MicroPython code known as a *library* — in this case, the **machine** library. The **machine** library contains all the instructions MicroPython needs to communicate with the Pico and other MicroPython-compatible devices, extending the language for physical computing. Without this line, you won't be able to control any of your Pico's GPIO pins — and you won't be able to make the on-board LED light up.

> ### ? SELECTIVE IMPORTS
>
> In both MicroPython and Python it's possible to import part of a library, rather than the whole library. This can use less memory and allows you to refer to functions without their library name prefix. Most programs in this book import whole libraries; elsewhere you may see programs with lines like `from machine import Pin`; this imports only the **Pin** function, rather than the whole **machine** library.

The **machine** library exposes what is known as an *application programming interface (API)*. The name sounds complicated, but describes exactly what it does: it provides a way for your program, or the *application*, to communicate with the Pico via an *interface*.

The next line of your program provides an example of the **machine** library's API:

```
led_onboard = machine.Pin("LED", machine.Pin.OUT)
```

This line defines an object called **led_onboard**, which offers a friendly name you can use to refer to the on-board LED later in your program. It's technically possible to use any name here, but it's best to stick with names which describe the variable's purpose, to make the program easier to read and understand.

The second part of the line calls the **Pin** function in the machine library. This function, as its name suggests, is designed for handling your Pico's GPIO pins. At the moment, none of the GPIO pins — including the on-board LED pin — know what they're supposed to be doing. The first argument, **"LED"**, tells the **Pin** function to use the GPIO assigned to the on-board LED, which means you don't need to remember its pin number. The second, **machine.Pin.OUT**, tells Pico the pin should be used as an *output* rather than an *input*.

That line alone is enough to set the pin up, but it won't light the LED. To do that, you need to tell your Pico to actually turn the pin on. Type the following code on the next line:

```
led_onboard.value(1)
```

This line is also using the machine library's API. Your earlier line created the object **led_onboard** as an output on the on-board LED pin; this line takes the object and sets its *value* to 1 for 'on'. It could also set the value to 0, for 'off'.

Click the **Run** button ▶ and save the program on your Pico as **Blink.py**. You'll see the LED light up. Congratulations: you've written your first physical computing program!

You'll notice, however, that the LED stays lit. That's because your program tells the Pico to turn it on, but never tells it to turn it off. You can add another line at the bottom of your program:

```
led_onboard.value(0)
```

Run the program this time, though, and the LED never seems to light up. That's because your Pico works very, very quickly — much faster than you can see with the naked eye. The LED is lighting up, but for such a short time that it appears to remain dark. To fix that, you need to slow your program down by introducing a delay.

Go back to the top of your program: click to move your cursor to the end of the first line and press **ENTER** to insert a new second line. On this line, type:

```
import time
```

Like **import machine**, this line imports a new library into MicroPython: the **time** library. This library handles everything to do with time, from measuring it to inserting delays into your programs.

Click on the end of the line **led_onboard.value(1)**, then press **ENTER** to insert a new line. Type:

```
time.sleep(5)
```

This calls the **sleep** function from the **time** library, which makes your program pause for the number of seconds you typed: in this case, five seconds.

Click the **Run** button again. This time you'll see the on-board LED on your Pico light up, stay lit for five seconds — try counting along — and go out again.

Finally, it's time to make the LED blink. To do that, you'll need to create a loop. Rewrite your program so it matches the one below:

```
import machine
import time

led_onboard = machine.Pin("LED", machine.Pin.OUT)

while True:
    led_onboard.value(1)
    time.sleep(5)
    led_onboard.value(0)
    time.sleep(5)
```

Remember that the lines inside the loop need to be indented by four spaces, so MicroPython knows they form the loop. Click the **Run** icon again, and you'll see the LED switch on for five seconds, switch off for five seconds, and switch on again, constantly repeating in an infinite loop. The LED will continue to flash until you click the **Stop** icon ⏹ to cancel your program and reset your Pico.

There's another way to handle the same job, too: using a *toggle*, rather than setting the LED's output to 0 or 1 explicitly. Delete the last four lines of your program and replace them so it looks like this:

```
import machine
import time

led_onboard = machine.Pin("LED", machine.Pin.OUT)

while True:
    led_onboard.toggle()
    time.sleep(5)
```

Run your program again. You'll see the same activity as before: the on-board LED will light up for five seconds, then go out for five seconds, then light up again in an infinite loop. This time, though, your program is two lines shorter: you've *optimised* it. Available on all digital output pins, **toggle()** simply switches between on and off: if the pin is currently on, **toggle()** switches it off; if it's off, **toggle()** switches it on.

CHALLENGE: LONGER LIGHT-UP

How would you change your program to make the LED stay on for longer? What about staying off for longer? What's the smallest delay you can use while still being able to see the LED blink on and off?

Using a breadboard

The next projects in this chapter will be much easier to complete if you use a solderless breadboard (**Figure 4-2**) to hold the components and make the electrical connections.

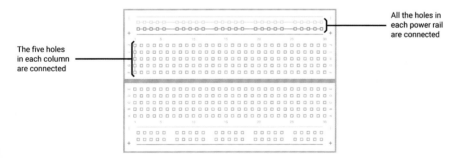

The five holes in each column are connected

All the holes in each power rail are connected

Figure 4-2   A solderless breadboard

A breadboard is covered with holes which are spaced 2.54mm apart to match most components. Under these holes are metal strips (*terminals*) which act like invisible jumper wires. These run in columns on the board, with most boards having a gap down the middle to split them in two halves. Many breadboards also have letters going up the left side and numbers on the top and bottom. These allow you to find a particular hole: A1 is the bottom-left, B1 is the hole just above it, while B2 is one hole to the right. A1 is connected to B1 by the hidden metal strips, but no number hole is ever connected to a different number hole unless you add a jumper wire.

Larger breadboards also have strips of holes along the top and bottom, typically marked with red and black or red and blue stripes. These are the *power rails*, and are designed to make wiring easier: you can connect a single wire from your Pico's ground pin to one of the power rails — typically marked with a blue or black stripe and a minus symbol — to provide a *common ground* for lots of components on the breadboard, and you can do the same if your circuit needs 3.3V or 5V power.

Adding electronic components to a breadboard is simple: just line their leads (the sticky-out metal parts) up with the holes and gently push until the component is in place. For connections you need to make beyond those the breadboard makes for you, you can use male-to-male (M2M) jumper wires; for connections from the breadboard to your Raspberry Pi Pico, use male-to-female (M2F) jumper wires.

Push your Pico into the breadboard so it straddles the middle gap and the micro USB port is at the edge of the board (see **Figure 4-3**). Pins 1 and 40 should be in the breadboard column marked with a 1, if your breadboard is numbered. Before pushing your Pico down, make sure the header pins are all properly positioned — if you bend a pin, it can be difficult to straighten it again without it breaking.

Gently push the Pico down until the plastic parts of the header pins are touching the breadboard. This means the metal parts of the header pins are fully inserted and making good electrical contact with the breadboard.

WARNING

Your Pico's pins are designed to be a fun and safe way to experiment with physical computing, but should always be treated with care. Be careful not to bend the pins, especially when you're inserting your Pico into a breadboard. Never connect two pins directly together, accidentally or deliberately, unless you're told to do so in a project's instructions: this is known as a *short circuit* and, depending on the pins, can permanently damage your Pico.

Figure 4-3 Your Pico is designed to sit securely in a solderless breadboard

Next steps: an external LED

So far, you've been working with your Pico on its own — running MicroPython programs on its RP2040 microcontroller and toggling the on-board LED on and off. Microcontrollers are usually used with *external* components, though — and your Pico is no exception.

For this project, you'll need a breadboard, male-to-male (M2M) jumper wires, an LED, and a 330 Ω resistor — or as close to 330 Ω as you have available. If you don't have a breadboard, you can use female-to-female (F2F) jumper wires, but the circuit will be fragile and easy to break.

RESISTANCE IS VITAL

The resistor is a vital component in this circuit: it protects your Raspberry Pi and the LED by limiting the amount of electrical current the LED can draw. Without it, the LED can pull too much current and burn itself — or your Raspberry Pi — out. When used like this, the resistor is known as a *current-limiting resistor*. The exact value of the resistor you need depends on the LED you're using, but 330 Ω works for most common LEDs. The higher the value, the dimmer the LED; the lower the value, the brighter the LED.

Never connect an LED to a Raspberry Pi without a current-limiting resistor, unless you know the LED has a built-in resistor of appropriate value.

Hold the LED in your fingers: you'll see one of its leads is longer than the other. The longer lead is known as the *anode*, and represents the positive side of the circuit; the shorter lead is the *cathode*, and represents the negative side. The anode needs to be connected to one of your Pico's GPIO pins via the resistor; the cathode needs to be connected to a ground pin.

With your Pico unplugged from USB, start by connecting the resistor: take either end and insert it into the breadboard in the same column as your Pico's GP15 pin at the bottom-right — if you're using a numbered breadboard with your Pico inserted at the edge, this should be column 20. Push the other end into a free column further down the breadboard — we're using column 26.

> ⚠ **WARNING**
>
> Never cram more than one component lead or jumper wire into a single hole on the breadboard. Remember: aside from the split in the middle, same-numbered holes are connected, so a component lead in A1 is connected to anything in B1, C1, D1, and E1.

Take the LED, and push the longer leg — the anode — into the same column as the end of the resistor. Push the shorter leg — the cathode — into the same column but across the centre gap in the breadboard, so it's lined up but not electrically connected to the longer leg except through the LED itself. Finally, insert a male-to-male (M2M) jumper wire into the same column as the shorter leg of the LED, then either connect it directly to one of your Pico's ground pins (via another hole in its column) or to the negative side of your breadboard's power rail. If you connect it to the power rail, finish the circuit by connecting the rail to one of your Pico's ground pins. Your finished circuit should look like **Figure 4-4**. Connect your Pico to your Raspberry Pi or computer.

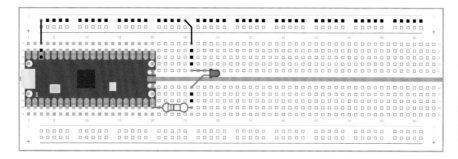

Figure 4-4   The finished circuit, with an LED and a resistor

Controlling an external LED in MicroPython is no different to controlling your Pico's internal LED: only the pin number changes. If you closed Thonny, reopen it and load your **Blink.py** program from earlier in the chapter. Find the line:

```
led_onboard = machine.Pin("LED", machine.Pin.OUT)
```

Edit the pin number, changing it from the string **"LED"** — the pin connected to your Pico's internal LED — to 15, the pin to which you connected the external LED. Also edit the name you created: you're not using the on-board LED anymore, so have it say **led_external** instead. You'll also have to change the name elsewhere in the program, until it looks like this:

```
import machine
import time

led_external = machine.Pin(15, machine.Pin.OUT)

while True:
    led_external.toggle()
    time.sleep(5)
```

PIN NUMBERS

The GPIO pins on your Pico are usually shown in pinout diagrams with their full names, such as GP15. In MicroPython, though, the letters G and P are dropped — so make sure you write **15** rather than **GP15** in your program or it won't work!

You don't really *need* to change the name in the program: it would run just the same if you'd left it at **led_onboard**, as it's only the pin number which truly matters. When you come back to the program later, though, it would be very confusing to have an object named **led_onboard** which lights up an external LED — try to get into the habit of making sure your names match their purpose!

CHALLENGE: MULTIPLE LEDS

Can you modify the program to light up both the on-board and external LEDs at the same time? Can you write a program which lights up the on-board LED when the external LED is switched off, and vice versa? Can you extend the circuit to include more than one external LED? Remember, you'll need a current-limiting resistor for every LED you use!

Inputs: reading a button

Outputs like LEDs are one thing, but the 'input/output' part of 'GPIO' means you can use pins as inputs too. For this project, you'll need a breadboard, male-to-male jumper wires, and a push-button switch. If you don't have a breadboard, you can use female-to-female (F2F) jumper wires, but the button will be much harder to press without accidentally breaking the circuit.

With your Pico unplugged from USB, remove any other components from your breadboard except your Pico, and begin by adding the push-button switch. If your push-button has only two legs, make sure they're in different-numbered columns on the breadboard somewhere to the right of you Pico. If it has four legs, turn it so the flat sides (the sides the legs *don't* stick out from) are aligned in the same numbered column, but also straddling the centre divide of the breadboard (as seen in **Figure 4-5**).

Connect the positive power rail of your breadboard to your Pico's 3V3 pin, and from there to one of the legs of the switch; then connect the other leg to pin GP14 on your Pico — it's the one just to the left of the pin you used for the LED project, and should be in column 19 of your breadboard.

If you're using a push-button with four legs, your circuit will only work if you use the correct pair of legs: the legs are connected in pairs, so you need to either use the two legs on the same side of the centre divide or diagonally opposite legs.

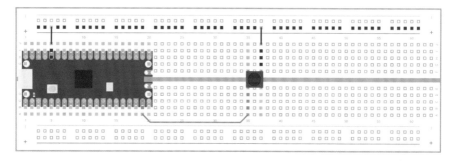

Figure 4-5   Wiring a four-leg push-button switch to GP14

Connect your Pico to USB again. Next, load Thonny, if you haven't already, and start a new program with the usual line:

```
import machine
```

Next, set up a pin as an input, rather than an output:

```
button = machine.Pin(14, machine.Pin.IN, machine.Pin.PULL_DOWN)
```

This works in the same way as your LED projects: an object called 'button' is created, which includes the pin number — GP14, in this case — and configures it as an input with the resistor set to pull-down. Creating the object, though, doesn't mean it will do anything by itself — just as creating the LED objects earlier didn't make the LEDs light up.

To actually read the button, you need to use the **machine** API again — this time using the **value** function to read, rather than set, the value of the pin. Type the following line:

```
print(button.value())
```

Click the **Run** icon and save your program as **Button.py** — remembering to make sure it saves on your Raspberry Pi Pico. Your program will print out a single number: the value of the input on GP14. Because the input is using a pull-down resistor, this value will be 0 — letting you know the button isn't pushed.

Hold down the button with your finger, and press the **Run** icon again. This time, you'll see the value 1 printed to the Shell: pushing the button has completed the circuit and changed the value read from the pin.

To read the button continuously, you'll need to add a loop to your program. Edit the program so it reads as below:

```
import machine
import time

button = machine.Pin(14, machine.Pin.IN, machine.Pin.PULL_DOWN)

while True:
    if button.value() == 1:
        print("You pressed the button!")
        time.sleep(2)
```

Click the **Run** button again. Nothing will happen until you press the button; when you do, you'll see a message printed to the Shell area. The delay, meanwhile, is important: your Pico runs a lot faster than you can read, and without the delay even a brief button press will print hundreds of messages!

You'll see the message print every time you press the button. If you hold the button down for longer than the two-second delay, it will print the message every two seconds until you let go of the button.

Inputs and outputs: putting it all together

Most circuits have more than one component, which is why your Pico has so many GPIO pins. It's time to put everything you've learned together to build a more complex circuit: a device which switches an LED on and off with a button.

This circuit combines the previous two, which used pin GP15 to drive the external LED, and GP14 to read the button; now rebuild your circuit so the LED and the button are on the breadboard at the same time, still connected to GP15 and GP14 (see **Figure 4-6**). Remember the LED's current-limiting resistor and to disconnect from USB while you're building the circuit!

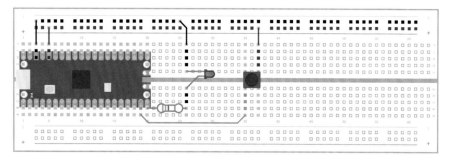

Figure 4-6   The finished circuit, with both a button and an LED

Start a new program in Thonny, import these two libraries:

```
import machine
import time
```

Next, set up both the input and output pins:

```
led_external = machine.Pin(15, machine.Pin.OUT)
button = machine.Pin(14, machine.Pin.IN, machine.Pin.PULL_DOWN)
```

Then create a loop which reads the button:

```
while True:
    if button.value() == 1:
```

Rather than printing a message to the Shell, this time you'll toggle the output pin (and the LED connected to it) based on the value of the input pin. Type the following, remembering it will need to be indented by eight spaces — which Thonny should have done when you pressed **ENTER** after the line above:

```
led_external.value(1)
time.sleep(2)
```

That's enough to turn the LED on, but you'll also need to turn it off again when the button isn't being pressed. Add the following new line, using the **BACK-SPACE** key to delete four of the eight spaces — meaning the line will not be part of the **if** statement, but will form part of the infinite loop:

```
led_external.value(0)
```

Your finished program should look like this:

```
import machine
import time

led_external = machine.Pin(15, machine.Pin.OUT)
button = machine.Pin(14, machine.Pin.IN, machine.Pin.PULL_DOWN)

while True:
    if button.value() == 1:
        led_external.value(1)
        time.sleep(2)
    led_external.value(0)
```

Save the program as **Switch.py** on your Pico and click **Run**. At first, nothing will happen; push the button, and you'll see the LED light up. Let go of the button; after two seconds, the LED will go out until you press the button again.

Congratulations: you've built your first circuit which controls one pin based on the input from another — a building block for bigger things!

CHALLENGE: BUILDING IT UP

Can you modify your program so it both lights the LED and prints a status message to the Shell? What would you need to change to make the LED stay on when the button isn't pressed and switch off when it is? Can you add more buttons and LEDs to the circuit?

WAIT

Chapter 5

Traffic light controller

Create your own mini pedestrian crossing system with multiple LEDs and a push-button

Microcontrollers can be found in almost all the electronic items you use on a daily basis — including traffic lights. A traffic light controller is a specially-built system which changes the lights on a timer, watches for pedestrians looking to cross, and can even adjust the timing of the lights depending on how much traffic there is — talking to nearby traffic light systems to ensure the whole traffic network keeps flowing smoothly.

While building a large-scale traffic management system is a pretty advanced project, it's simplicity itself to build a miniature simulator powered by your Raspberry Pi Pico. With this project, you'll see how to control multiple LEDs, set different timings, and how to monitor a push-button input while the rest of the program continues to run using a technique known as *threading*.

For this project, you'll need your Pico; a breadboard; a red, yellow (or amber), and green LED; three 330 Ω resistors; an active piezoelectric buzzer; and a selection of male-to-male (M2M) jumper wires. You'll also need a micro USB cable to connect your Pico to your Raspberry Pi or other computer running Thonny.

A simple traffic light

Disconnect your Pico from USB, and build the traffic light system shown in **Figure 5-1**. Take your red LED and insert it into the breadboard so it straddles the centre divide. Use one 330 Ω resistor, and a jumper wire if you need to make a longer connection, to connect the longer leg — the anode — of the LED to the pin at the bottom-right of your Pico as seen from the top with the micro USB cable leftmost, GP15. If you're using a numbered breadboard and have your Pico inserted as shown, this will be column 20.

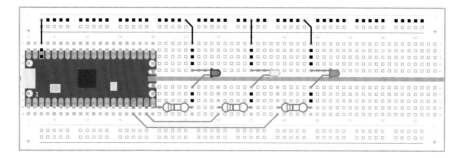

Figure 5-1 A basic three-light traffic light system

WARNING

Always remember that an LED needs a current-limiting resistor before it can be connected to your Pico. Without it, the best outcome is the LED will burn out and no longer work; the worst outcome is it could do the same to your Pico.

Take a jumper wire and connect the shorter leg — the cathode — of the red LED to your breadboard's ground rail. Take another, and connect the ground rail to one of your Pico's ground (GND) pins — in **Figure 5-1**, we've used the ground pin on column three of the breadboard.

You've now got one LED connected to your Pico, but a real traffic light has at least three in all: a red light to tell the traffic to stop, amber or yellow to tell the traffic the light is about to change, and green to tell the traffic it can go again.

Take your amber or yellow LED and wire it to your Pico in the same way as the red LED, making sure the shorter leg connected to the ground rail of the breadboard. This time, though, wire the longer leg — via the 330 Ω resistor — to the pin next to the one to which you wired the red LED, GP14.

Finally, take the green LED and wire it up the same way again — remembering the 330 Ω resistor — to pin GP13. This isn't the pin right next to pin GP14, though — that pin is a ground (GND) pin, which you can see if you look closely at your Pico: the ground pins all have a square shape to their pads, while the other pins are round.

When you've finished, your circuit should match **Figure 5-1**: a red, a yellow or amber, and a green LED, all wired to different GPIO pins on your Pico via individual 330 Ω resistors and connected to a shared ground pin via your breadboard's ground rail.

To program your traffic lights, connect your Pico to your Raspberry Pi or other computer and load Thonny. Create a new program, and start by importing the `machine` library so you can control your Pico's GPIO pins:

```
import machine
```

You'll also need to import the **time** library, so you can add delays between the lights going on and off:

```
import time
```

As with any program using your Pico's GPIO pins, you'll need to set each pin up before you can control it:

```
led_red = machine.Pin(15, machine.Pin.OUT)
led_amber = machine.Pin(14, machine.Pin.OUT)
led_green = machine.Pin(13, machine.Pin.OUT)
```

These lines set pins GP15, GP14, and GP13 up as outputs, and each is given a descriptive name to make it easier to read the code: **led**, so you know the pins control an LED, and then the colour of the LED.

Real traffic lights don't run through once and stop — they keep going, even when there's no traffic there and everyone's asleep. So that your program does the same, you'll need to set up an infinite loop:

```
while True:
```

You'll need to indent all the lines beneath this by four spaces, so MicroPython knows they form part of the loop; when you press the **ENTER** key Thonny will automatically indent the lines for you.

```
    led_red.value(1)
    time.sleep(5)
    led_amber.value(1)
    time.sleep(2)
    led_red.value(0)
    led_amber.value(0)
    led_green.value(1)
    time.sleep(5)
    led_green.value(0)
    led_amber.value(1)
    time.sleep(5)
    led_amber.value(0)
```

Save the program to your Pico as **Traffic_Lights.py** and click the **Run** icon. Watch the LEDs: red lights up first, telling traffic to stop; next, amber comes on to warn drivers the lights are about to change; then both switch off and green comes on to let traffic know it can pass; then green turns off and amber comes on to warn drivers the lights are about to change again; finally, amber turns off — and the loop restarts from the beginning, with red coming on.

The pattern will loop until you press the Stop button, because it forms an infinite loop. It's based on the traffic light pattern used in real-world traffic control systems in the UK and Ireland, but sped up — giving cars just five seconds to pass through the lights wouldn't let the traffic flow very freely!

Real traffic lights aren't just there for road vehicles, though: they are also there to protect pedestrians, giving them an opportunity to cross a busy road safely. In the UK, the most common type of these lights are known as *pedestrian-operated user-friendly intelligent crossings* or *puffin crossings*.

To turn your traffic lights into a puffin crossing, you'll need two things: a push-button switch, so the pedestrian can ask the lights to let them cross the road; and a buzzer, so the pedestrian knows when it's their turn to cross. Wire those into your breadboard as in **Figure 5-2**, with the switch wired to pin GP16 and the 3V3 rail of your breadboard, and the buzzer wired to pin GP12 and the breadboard's ground rail. Disconnect the Pico from USB while you build this.

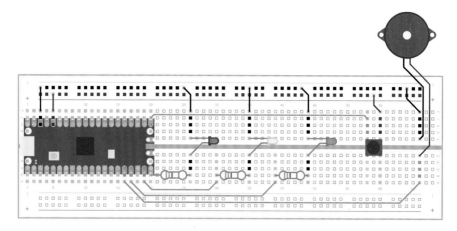

Figure 5-2　A puffin crossing traffic light system

If you run your program again, you'll find the button and buzzer do nothing. That's because you haven't yet told your program how to use them. In Thonny, go back to the lines where you initialised your LEDs and add two lines below:

```
button = machine.Pin(16, machine.Pin.IN, machine.Pin.PULL_DOWN)
buzzer = machine.Pin(12, machine.Pin.OUT)
```

This sets the button on pin GP16 up as an input, and the buzzer on pin GP12 as an output. Remember, your Raspberry Pi Pico has built-in programmable resistors for its inputs, which we are setting to pull-down mode for the projects in this book. This means that the pin's voltage is pulled down to 0 V (and its logic level is 0), unless it is connected to 3.3V power (in which case its logic level will be 1 until disconnected).

Next, you need a way for your program to constantly monitor the value of the button. Previously, all your programs have worked step-by-step through a list of instructions — only ever doing one thing at a time. Your traffic light program is no different: as it runs, MicroPython walks through your instructions step-by-step, turning the LEDs on and off.

For a basic set of traffic lights, that's enough; for a puffin crossing, though, your program needs to be able to record whether the button has been pressed in a way that doesn't interrupt the traffic lights. To make that work, you'll need a new library: **_thread**. Go back to the section of your program where you import the machine and time libraries, and import the **_thread** library:

```
import _thread
```

A *thread* or *thread of execution* is, effectively, a small and partially independent program. You can think of the loop you wrote earlier, which controls the lights, as the *main thread* of your program — and using the **_thread** library you can create an additional thread, running at the same time.

An easy way to visualise threads is to think of each one as a separate worker in a kitchen: while the chef is preparing the main dish, someone else is working on a sauce. At the moment, your program has only one thread — the one which controls the traffic lights. The RP2040 microcontroller which powers your Pico, however, has two processing cores — meaning, like the chef and the sous chef in the kitchen, you can run two threads at the same time to get more work done.

Before you can make another thread, you'll need a way for the new thread to share information with the main thread — you can do this with *global variables*. The variables you've been working with prior to this are known as *local variables*, and only work in one section of your program; a global variable works everywhere, meaning one function (including one running inside a thread) can change the value and another can check to see if it has been changed.

To start, create a global variable. Below your **buzzer =** line, add the following:

```
global button_pressed
button_pressed = False
```

This sets up **button_pressed** as a global variable, and gives it a default value of **False** — meaning when the program starts, the button hasn't yet been pushed. The next step is to define what you want to have happen in your thread, by adding the following lines directly below — adding a blank line, if you want, to make your program more readable:

```
def button_reader_thread():
    global button_pressed
    while True:
        if button.value() == 1:
            button_pressed = True
        time.sleep(0.01)
```

The first line you've added defines a function and gives it a descriptive name: it's a function to read the button input, and is intended to be run within a thread. Like when writing a loop, MicroPython needs everything contained within the function to be indented by four spaces — so it knows where the function begins and ends.

The next line lets MicroPython know you need to change the value of the global **button_pressed** variable. If you only want to check the value, you wouldn't need this line — but without it you can't make any changes to the variable.

Next, you've set up a new loop — which means a new four-space indent needs to follow, for eight in total, so MicroPython knows both that the loop is part of the function and the code below is part of the loop. This nesting of code in multiple levels of indentation is very common in MicroPython, and Thonny will do its best to help you by automatically adding a new level each time it's needed — but it's up to you to remember to delete the spaces it adds when you're finished with a particular section of the program.

The next line is a conditional which checks to see if the value of the button is 1. Because your Pico is using an internal pull-down resistor, when the button isn't being pressed the value read is 0 — meaning the code under the conditional never runs. Only when the button is pressed will the final line of your thread run: a line which sets the **button_pressed** variable to **True**, letting the rest of your program know the button has been pushed. Finally, we add a very short (0.01 second) delay to prevent the **while** loop running too fast.

You might notice there's nothing in the thread to reset the **button_pressed** variable back to **False** when the button is released after being pushed. There's a reason for that: while you can push the button of a puffin crossing at any time during the traffic light cycle, it only takes effect when the light has gone red and it's safe for you to cross. All that the function running inside the new thread needs to do is to change the variable when the button has been pushed; the code in your main thread will handle resetting it back to **False** when the pedestrian has safely crossed the road.

Right now, **button_reader_thread** is just an ordinary function with the word **thread** in its name. We haven't done anything special to make it a thread, so there's nothing to set it running: it's possible to start a thread at any point in your program, and you'll need to specifically tell the **_thread** library when you want to launch the thread. Unlike running a normal line of code, running the thread doesn't pause the rest of the program: when the thread starts, MicroPython will carry on and run the next line of your program even as it runs the first line of your new thread.

You need to add something to launch that function within a thread. Create a new line below your function (deleting all the indentation Thonny has automatically added for you) which reads:

```
_thread.start_new_thread(button_reader_thread, ())
```

This tells the **_thread** library to start the function you defined earlier. At this point, the thread will start to run and quickly enter its loop — checking the button thousands of times a second to see if it's been pressed yet. The main thread, meanwhile, will carry on with the main part of your program.

Click the **Run** button now. You'll see the traffic lights carry on their pattern exactly as before, with no delay or pauses. If you press the button, though, nothing will happen — because you haven't added the code to actually react to the button yet.

Go to the start of your main loop, and add the following code directly underneath the line **while True:** — remembering to pay attention to the nested indentation, and deleting the indentation Thonny has added when it's no longer required:

```
if button_pressed == True:
    led_red.value(1)
    for i in range(10):
        buzzer.value(1)
        time.sleep(0.04)
        buzzer.value(0)
        time.sleep(0.2)
    global button_pressed
    button_pressed = False
```

This chunk of code checks the **button_pressed** global variable to see if the push-button switch has been pressed at any time since the loop last ran. If it has, as reported by the button reading thread you made earlier, it begins running a section of code which starts by turning the red LED on to stop traffic and then beeps the buzzer ten times — letting the pedestrian know it's time to cross.

Finally, the last two lines reset the **button_pressed** variable back to **False** — so the next time the loop runs it won't trigger the pedestrian crossing code unless the button has been pushed again. You'll see you didn't need the line **global button_pressed** to check the status of the variable in the conditional; it's only needed when you want to change the variable and have that change affect other parts of your program.

Your program should look like this:

```
import machine
import time
import _thread

led_red = machine.Pin(15, machine.Pin.OUT)
led_amber = machine.Pin(14, machine.Pin.OUT)
led_green = machine.Pin(13, machine.Pin.OUT)
button = machine.Pin(16, machine.Pin.IN, machine.Pin.PULL_DOWN)
buzzer = machine.Pin(12, machine.Pin.OUT)

global button_pressed
button_pressed = False

def button_reader_thread():
    global button_pressed
    while True:
        if button.value() == 1:
            button_pressed = True
        time.sleep(0.01)

_thread.start_new_thread(button_reader_thread, ())

while True:
    if button_pressed == True:
        led_red.value(1)
        for i in range(10):
            buzzer.value(1)
            time.sleep(0.04)
            buzzer.value(0)
            time.sleep(0.2)
        global button_pressed
        button_pressed = False
    led_red.value(1)
```

```
time.sleep(5)
led_amber.value(1)
time.sleep(2)
led_red.value(0)
led_amber.value(0)
led_green.value(1)
time.sleep(5)
led_green.value(0)
led_amber.value(1)
time.sleep(5)
led_amber.value(0)
```

Click the **Run** icon. At first, the program will run as normal: the traffic lights will go on and off in the usual pattern. Press the push-button switch: if the program is currently in the middle of its loop, nothing will happen until it reaches the end and loops back around again — at which point the light will go red and the buzzer will beep to let you know it's safe to cross the road.

The conditional section of code for crossing the road runs before the code you wrote earlier for turning the lights on and off in a cyclic pattern: after it's finished, the pattern will begin as usual with the red LED staying lit for a further five seconds on top of the time it was lit while the buzzer was going. This mimics how a real puffin crossing works: the red light remains lit even after the buzzer has stopped sounding, so anyone who started to cross the road while the buzzer was going has time to reach the other side before the traffic is allowed to go.

Let the traffic lights loop through their cycle a few more times, then press the button again to trigger another crossing. Congratulations: you've built your own puffin crossing!

CHALLENGE: CAN YOU IMPROVE IT?

Can you change the program to give the pedestrian longer to cross? Can you find information about other countries' traffic light patterns and reprogram your lights to match? Can you add a second button, so the pedestrian on the other side of the road can signal they want to cross too?

Chapter 6

Reaction game

Build a simple reaction timing game using an LED and push-buttons, for one or two players

Microcontrollers aren't only found in industrial devices: they power plenty of electronics around the home, including toys and games. In this chapter you're going to build a simple reaction timing game, seeing who among your friends will be the first to press a button when a light goes off.

The study of reaction time is known as *mental chronometry* and while it forms a hard science, it is also the basis of plenty of skill-based games — including the one you're about to build. Your reaction time — the time it takes your brain to process the need to do something and send the signals to make that something happen — is measured in milliseconds: the average human reaction time is around 200–250 milliseconds, though some people enjoy considerably faster reaction times that will give them a real edge in the game!

For this project you'll need your Pico; a breadboard; an LED of any colour; a single 330 Ω resistor; two push-button switches; and a selection of male-to-male (M2M) jumper wires. You'll also need a micro USB cable to connect your Pico to your Raspberry Pi or other computer running Thonny.

A single-player game

With your Pico inserted in your breadboard, but not plugged into USB, start by placing your LED into your breadboard so that it straddles the centre divide. Remember that LEDs only work when they're the right way around: make sure you identify which is the longer leg, or the anode, and which is the shorter leg, the cathode.

Using a 330 Ω current-limiting resistor, to protect both the LED and your Pico, wire the longer leg of the LED to pin GP15 at the bottom-right of your Pico as oriented when the micro USB cable is to the left. If you're using a numbered breadboard and have your Pico inserted as shown in **Figure 6-1**, this will be column 20.

> **WARNING**
>
> It bears repeating that an LED always needs a current-limiting resistor before it can be connected to your Pico. Without the resistor, the LED could burn out — or your Pico could be damaged.

Take a jumper wire and connect the shorter leg of the LED to your breadboard's ground rail. Take another and connect the ground rail to one of your Pico's ground (GND) pins — in **Figure 6-1**, we've used the ground pin on column three of the breadboard.

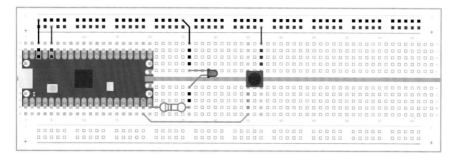

Figure 6-1   A single-player reaction game

Next, add the push-button switch as shown in **Figure 6-1**. Take a jumper wire and connect one of the push-button's switches to pin GP14, right next to the pin you used for your LED. Use another jumper wire to connect the other leg — the one diagonally opposite the first, if you're using a four-leg push-button switch — to your breadboard's power rail. Finally, take a last jumper wire and connect the power rail to your Pico's 3V3 pin.

> **WHY 3V3?**
>
> Remember that switches, like LEDs, need resistors to operate correctly, and that your Pico has programmable resistors on all its GPIO pins. In the projects for this book, we are setting them to pull-down resistors, meaning the pin has to be pulled high when the push-button switch is pressed — which is what wiring the switch to the 3V3 pin via the breadboard's power rail does.

Your circuit now has everything it needs to act as a simple single-player game: the LED is the output device, taking the place of the TV you would normally use with a games console; the push-button switch is the controller; and your Pico is the games console, albeit one considerably smaller than you'd usually see!

Now you need to write the game. As before, connect your Pico to your Raspberry Pi or other computer and load Thonny. Create a new program, and start it by importing the **machine** library so you can control your Pico's GPIO pins:

```
import machine
```

You're also going to need the **time** library. In addition, you'll need one more library: **random**, which handles creating random numbers — a key part of making a game fun, and used in this game to prevent a player who has played it before from simply counting down a fixed number of seconds from clicking the **Run** button.

```
import time
import random
```

Next, set a **pressed** variable to False (more on this later) and set up the two pins you're using: GP15 for the LED, and GP14 for the push-button switch.

```
pressed = False
led = machine.Pin(15, machine.Pin.OUT)
button = machine.Pin(14, machine.Pin.IN, machine.Pin.PULL_DOWN)
```

In previous chapters, you've handled push-button switches in either the main program or in a separate thread. This time, though, you're going to take a different and more flexible approach: *interrupt requests*, or *IRQs*.

The name sounds complex, but it's simple: imagine you're reading a book, page by page, and someone comes up to you and asks you a question. That person is performing an interrupt request: asking you to stop what you're doing, answer their question, then letting you go back to reading your book.

A MicroPython interrupt request works in the same way: it allows something, in this case the press of a push-button switch, to interrupt the main program. In some ways it's similar to a thread, in that there's a chunk of code which sits outside the main program. Unlike a thread, though, the code isn't constantly running: it only runs when the interrupt is triggered.

Start by defining a *handler* (also known as a *callback function*) for the interrupt. This code runs when the interrupt is triggered. As with any kind of nested code, the handler's code — everything after the first line — must be indented by four spaces for each level; Thonny will do this for you automatically.

```python
def btn_handler(pin):
    global pressed
    if not pressed:
        pressed = True
        print(pin)
```

This handler checks the status of the **pressed** variable and sets it to **True** to ignore further button presses (thus ending the game). It then prints out information about the pin responsible for triggering the interrupt. That's not too important at the moment — you only have one pin configured as an input, GP14, so the interrupt will always come from that pin — but lets you test your interrupt easily.

Continue your program below, remembering to delete the indent that Thonny has automatically created — the following code is not part of the handler:

```python
led.value(1)
time.sleep(random.uniform(5, 10))
led.value(0)
```

This code will be immediately familiar to you: the first line turns on the LED that's connected to pin GP15; the next line pauses the program; the last line turns the LED off again — the player's signal to push the button. Rather than using a fixed delay, however, it makes use of the **random** library to pause the program for between five and ten seconds — the 'uniform' part referring to a uniform distribution between those two numbers.

At the moment, though, there's nothing watching for the button being pushed. You need to set up the interrupt for that, by typing in the following line at the bottom of your program:

```python
button.irq(trigger=machine.Pin.IRQ_RISING, handler=btn_handler)
```

Setting up an interrupt requires two things: a *trigger* and a *handler*. The trigger tells your Pico what it should be looking for as a valid signal to interrupt what it's doing; the handler, which you defined earlier in your program, is the code which runs after the interrupt is triggered.

In this program your trigger is **IRQ_RISING**: this triggers the interrupt when the pin's value rises from low, its default state thanks to the built-in pull-down resistor, to high, when the button connected to 3V3 is pushed. A trigger of **IRQ_FALLING** would do the opposite: trigger the interrupt when the pin goes from high to low. In the case of your circuit, **IRQ_RISING** will trigger as soon as the button is pushed; **IRQ_FALLING** would trigger only when the button is released.

THE RISE AND FALL OF IRQS

If you need to write a program which triggers an interrupt whenever a pin changes, without caring whether it's rising or falling, you can combine the two triggers using a pipe or vertical bar symbol (**|**):

```
button.irq(trigger=machine.Pin.IRQ_RISING |
machine.Pin.IRQ_FALLING, handler=btn_handler)
```

Your program should look like this:

```
import machine
import time
import random

pressed = False
led = machine.Pin(15, machine.Pin.OUT)
button = machine.Pin(14, machine.Pin.IN, machine.Pin.PULL_DOWN)

def btn_handler(pin):
    global pressed
    if not pressed:
        pressed = True
        print(pin)

led.value(1)
time.sleep(random.uniform(5, 10))
led.value(0)
button.irq(trigger=machine.Pin.IRQ_RISING, handler=btn_handler)
```

Click **Run** and save the program to your Pico as **Reaction_Game.py**. You'll see the LED light up: that's your signal to get ready with your finger on the button. When the LED goes out, press the button as quickly as you can.

When you press the button, it triggers the handler code you wrote earlier. Look at the Shell area: you'll see your Pico has printed a message, confirming that the interrupt was triggered by pin GP14. You'll also see another detail: **mode=IN** tells you the pin was configured as an input.

That message doesn't make for much of a game, though: for that, you need a way to time the player's reaction speed. Start by deleting the line **print(pin)**

---

from your button handler — you don't need it. Add this new line just above the call to **button.irq()**:

```
start_time = time.ticks_ms()
```

This creates a new variable called **start_time** and fills it with the output of the **time.ticks_ms()** function, which counts the number of milliseconds that have elapsed since the **time** library began counting. This provides a reference point: the time just after the LED went out and just before the interrupt trigger became ready to read the button press.

Next, go back to your button handler and add the following two lines after **pressed = True**, remembering that they'll need to be indented by eight spaces so MicroPython knows they form part of the nested code:

```
react_time = time.ticks_diff(time.ticks_ms(), start_time)
print(f"Your reaction time: {react_time} milliseconds!")
```

The first line creates another variable to track when the interrupt was triggered (in other words, when you pressed the button). Rather than simply taking a reading from **time.ticks_ms()** as before, it uses **time.ticks_diff()** — a function that provides the difference between when this line of code is triggered and the reference point held in the variable **start_time**.

The second line prints the result with a *formatted string* (*f-string*) to print it nicely. The **f** before the opening **"** indicates the start of an f-string. When MicroPython encounters an expression in an f-string within curly brackets (**{** and **}**), it inserts the value into the string. That is the **react_time** variable in this case — the difference, in milliseconds, between when you took the reference point for the timer and when the button triggered the interrupt.

Your program should now look like this:

```
import machine
import time
import random

pressed = False
led = machine.Pin(15, machine.Pin.OUT)
button = machine.Pin(14, machine.Pin.IN, machine.Pin.PULL_DOWN)

def btn_handler(pin):
    global pressed
    if not pressed:
        pressed = True
```

```
    react_time = time.ticks_diff(time.ticks_ms(), start_time)
    print(f"Your reaction time: {react_time} milliseconds!")

led.value(1)
time.sleep(random.uniform(5, 10))
led.value(0)
start_time = time.ticks_ms()
button.irq(trigger=machine.Pin.IRQ_RISING, handler=btn_handler)
```

Click the **Run** button, wait for the LED to go out, and push the button. This time, instead of a report on the pin that triggered the interrupt, you'll see a line telling you how quickly you pushed the button — a measurement of your reaction time. Click the **Run** button again and see if you can push the button more quickly this time — in this game, you're trying for as low a score as possible!

CHALLENGE: CUSTOMISATION

Can you tweak your game so that the LED stays lit for a longer time? What about staying lit for a shorter time? Can you personalise the message that prints to the Shell area, and add a second message congratulating the player?

A two-player game

Single-player games are fun, but getting your friends involved is even better. You can start by inviting them to play your game and comparing your high — or, rather, low — scores to see who has the quickest reaction time. Then, you can modify your game to let you go head-to-head!

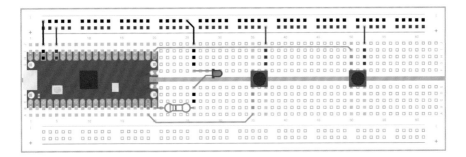

Figure 6-2   The circuit for a two-player reaction game

Start by adding a second button to your circuit (disconnect the Pico from USB first). Wire it like the first button, with one leg going to the breadboard power rail and the other to pin GP16 — the pin across the board from GP15 where the LED is connected, at the opposite corner of your Pico as shown in **Figure 6-2**.

Make sure the two buttons are spaced far enough apart that each player has room to put their finger on their button.

Connect the Pico to your Raspberry Pi or other computer. Although your second button is now connected to your Pico, it doesn't know what to do with it yet. Go back to your program in Thonny and find where you set up the first button. Directly beneath this line, add:

```
right_btn = machine.Pin(16, machine.Pin.IN, machine.Pin.PULL_DOWN)
```

You'll notice that the name now specifies which button you're working with: the right-hand button. To avoid confusion, edit the line above so that you make it clear that the first button you connected is now the left-hand button:

```
left_btn = machine.Pin(14, machine.Pin.IN, machine.Pin.PULL_DOWN)
```

You'll need to make the same change elsewhere in your program, too. Scroll to the bottom of your code and change the line that sets up the interrupt trigger to:

```
left_btn.irq(trigger=machine.Pin.IRQ_RISING, handler=btn_handler)
```

Add another line beneath it to set up an interrupt trigger on your new button:

```
right_btn.irq(trigger=machine.Pin.IRQ_RISING, handler=btn_handler)
```

Your program should now look like this:

```
import machine
import time
import random

pressed = False
led = machine.Pin(15, machine.Pin.OUT)
left_btn = machine.Pin(14, machine.Pin.IN, machine.Pin.PULL_DOWN)
right_btn = machine.Pin(16, machine.Pin.IN, machine.Pin.PULL_DOWN)

def btn_handler(pin):
    global pressed
    if not pressed:
        pressed = True
        react_time = time.ticks_diff(time.ticks_ms(), start_time)
        print(f"Your reaction time: {react_time} milliseconds!")

led.value(1)
time.sleep(random.uniform(5, 10))
```

```
led.value(0)
start_time = time.ticks_ms()
left_btn.irq(trigger=machine.Pin.IRQ_RISING, handler=btn_handler)
right_btn.irq(trigger=machine.Pin.IRQ_RISING, handler=btn_handler)
```

Click the **Run** icon, wait for the LED to go out, then press the left-hand push-button switch: you'll see that the game works the same as before, printing your reaction time to the Shell area. Click the **Run** icon again, but this time when the LED goes out, press the right-hand button: the game will work just the same, printing your reaction time as normal.

INTERRUPTS AND HANDLERS

Each interrupt you create needs a handler, but a single handler can deal with as many interrupts as you like. In the case of this program, you have two interrupts both going to the same handler — meaning that whichever interrupt triggers, they'll run the same code. A different program might have two handlers, letting each interrupt run different code — it all depends on what you need your program to do.

To make the game a little more exciting, you can have it report on which of the two players pressed the button first. Go back to the top of your program, just below where you initialised the LED and two buttons, and add the following:

```
fastest_btn = None
```

This sets up a new variable, **fastest_btn**, and sets its initial value to **None** (meaning no button has yet been pressed). Next, go to the bottom of your button handler and delete the two lines which handle the timer and printing, then replace them with:

```
global fastest_btn
fastest_btn = pin
```

Remember that these lines will need to be indented by eight spaces so that MicroPython knows they're part of the function. These two lines allow your function to change, rather than just read, the **fastest_btn** variable, and set it to contain the details of the pin which triggered the interrupt — the same details your game printed to the Shell area earlier in the chapter, including the number of the triggering pin.

Now go right to the bottom of your program, and add these two new lines:

```
while fastest_btn is None:
    time.sleep(1)
```

This creates a loop, but it's not an infinite loop: here, you've told MicroPython to run the code in the loop only when the **fastest_btn** variable is still zero (the value it was initialised with at the start of the program). In effect, this pauses your program's main thread until the interrupt handler changes the value of the variable. If neither player presses a button, the program will simply pause.

Finally, you need a way to determine which player won — and to congratulate them. Type the following at the bottom of the program, making sure to delete the four-space indent Thonny will have created for you on the first line — these lines do not form part of the loop:

```
if fastest_btn is left_btn:
    print("Left Player wins!")
elif fastest_btn is right_btn:
    print("Right Player wins!")
```

The first line sets up an **if** conditional which looks to see if the **fastest_btn** variable is **left_btn** — meaning the IRQ was triggered by the left-hand button. If so, it will print a message — with the line below indented by four spaces so that MicroPython knows it should run it only if the conditional is true — congratulating the left-hand player, whose button is connected to GP14.

The next line, which should not be indented, extends the conditional as an **elif** — short for 'else if', a way of saying 'if the first conditional wasn't true, check this conditional next'. This time it looks to see if the **fastest_btn** variable is **right_btn** — and, if so, prints a message congratulating the right-hand player, whose button is connected to GP16.

Your finished program should look like this:

```
import machine
import time
import random

pressed = False
led = machine.Pin(15, machine.Pin.OUT)
left_btn = machine.Pin(14, machine.Pin.IN, machine.Pin.PULL_DOWN)
right_btn = machine.Pin(16, machine.Pin.IN, machine.Pin.PULL_DOWN)
fastest_btn = None

def btn_handler(pin):
    global pressed
    if not pressed:
        pressed = True
        global fastest_btn
```

```
        fastest_btn = pin

led.value(1)
time.sleep(random.uniform(5, 10))
led.value(0)
start_time = time.ticks_ms()
left_btn.irq(trigger=machine.Pin.IRQ_RISING, handler=btn_handler)
right_btn.irq(trigger=machine.Pin.IRQ_RISING, handler=btn_handler)
while fastest_btn is None:
    time.sleep(1)
if fastest_btn is left_btn:
    print("Left Player wins!")
elif fastest_btn is right_btn:
    print("Right Player wins!")
```

Press the **Run** button and wait for the LED to go out — but don't press either of the push-button switches just yet. You'll see that the Shell area remains blank, and doesn't bring back the **>>>** prompt; that's because the main thread is still running, sitting in the loop you created.

Now push the left-hand button, connected to pin GP14. You'll see a message congratulating you printed to the Shell — your left hand was the winner! Click **Run** again and try pushing the right-hand button after the LED goes out: you'll see another message printed, this time congratulating your right hand. Click **Run** again, this time with one finger on each button: push them both at the same time and see whether your right hand or left hand is faster!

Now that you've created a two-player game, you can invite your friends to play along and see which of you has the fastest reaction times!

CHALLENGE: TIMINGS

Can you modify the messages that print? Can you add a third button, so that three people can play at once? Is there an upper limit to how many buttons you could add? Can you add the timer back into your program, so it tells the winning player how quick their reaction time was?

ALARM!

Chapter 7

Burglar alarm

Use a motion sensor to detect intruders and sound the alarm with a flashing light and siren

Another real-world use of microcontrollers is in alarm systems. From the alarm clock that gets you up in the morning to fire alarms, burglar alarms, and even the alarms that sound when there's a problem at a nuclear power station, microcontrollers help keep us all safe.

In this chapter you're going to build your own burglar alarm, which works in exactly the same way as a commercial version: a special motion sensor keeps watch for anyone entering the room who shouldn't be there, and flashes a light while sounding a siren to alert people to the intrusion. Whether you're protecting a bank vault or trying to keep spying siblings out of your room, or co-workers out of your cubicle, a burglar alarm is sure to come in handy.

For this project you'll need your Pico; a breadboard; an LED of any colour; a 330 Ω resistor; an active piezoelectric buzzer; one or more HC-SR501 passive infrared (PIR) sensors; and a selection of male-to-male (M2M) and male-to-female (M2F) jumper wires. You'll also need a micro USB cable to connect your Pico to a Raspberry Pi or computer running the Thonny MicroPython IDE.

The HC-SR501 PIR sensor

In previous chapters, you've been working with simple input components in the form of push-button switches. This time, you're going to be using a spe-cialised input known as a *passive infrared sensor* or *PIR*. There are hundreds of different PIR sensors available; the HC-SR501 is low-cost, high-performance, and works perfectly with your Pico.

A passive infrared sensor is designed to detect movement — in particular movement from people and other living things. It works a little like a camera, but instead of capturing visible light it looks for the heat emitted from a living body as infrared radiation. It's known as a passive infrared sensor, rather than active. That's because, like a camera sensor, it doesn't send out any signals of its own.

The actual sensor is buried underneath a plastic lens, typically shaped like a half-ball. The lens isn't technically necessary for the sensor to work, but serves to provide a wider *field of vision* (FOV); without the lens, the PIR sensor would only be able to see movement in a very narrow angle directly in front of the sensor. The lens serves to pull in infrared from a much wider angle, so a single PIR sensor can watch for movement over most of a room.

In commercial burglar alarm systems, a PIR sensor is only one of the sensors used; others include break-glass sensors which can tell when a window has been smashed, magnetic sensors which monitor whether a door is open or closed, acoustic sensors which can pick up a burglar's footsteps, and vibration sensors for telling if a lock is being forced open. A simple PIR sensor, though, is often enough for a low-security area — think a reception room, rather than a bank vault.

Pick up your HC-SR501 sensor now and look at it. The first thing to notice is that it has a circuit board of its own, a lot like your Pico — only smaller. As well as the sensor and lens, there are several other components: a small black *integrated circuit* (IC) which drives the sensor, some *capacitors*, and tiny surface-mount resistors. You may see one or more small *potentiometers* which you can twist with a screwdriver to adjust the sensitivity of the sensor and how long it stays active when triggered; leave these as they are for now.

You'll also see three male pins, exactly like the pins on the bottom of your Pico. You may not be able to push these directly into your breadboard, though, as the components on the board may get in the way. Instead of plugging it into the breadboard, take three male-to-female (M2F) jumper wires and insert the female ends onto the pins on your HC-SR501.

Next, be sure that your Pico is disconnected from USB. Then, take the male ends and wire them to the breadboard and your Pico. You'll need to check the documentation for your sensor when wiring it to your Pico: a lot of different companies make HC-SR501 sensors, and they don't always use the same pin configuration. For the sensor illustrated in **Figure 7-1**, the pins are set up so that the ground (GND) pin is on the left, the signal or trigger pin is in the middle, and the power pin is on the right; your sensor may need the wires placed in a different order!

Start with the ground wire: this needs to be connected to any of your Pico's ground pins. In **Figure 7-1**, it's connected to the first ground pin in the bottom row of breadboard column 3. Next, connect the signal wire: this should be connected to your Pico's GPIO pin GP28 (column 7 in **Figure 7-1**).

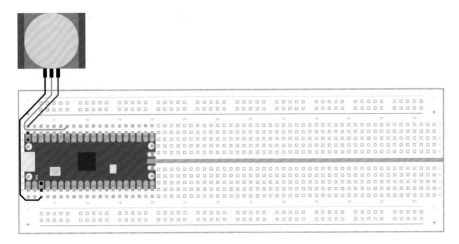

Figure 7-1   Wiring an HC-SR501 PIR sensor to your Pico

Finally, you need to connect the power wire. Don't connect the HC-SR501 to your Pico's 3V3 pin, though: the HC-SR501 is a 5 V device, meaning that it needs five volts of electricity to work. If you wire the sensor to your Pico's 3V3 pin, it won't work — the pin simply doesn't provide enough power.

To give your sensor the 5 V power it needs, wire it to pin 40 on your Pico — VBUS (if your Pico is oriented as shown in **Figure 7-1**, it's the top-left pin). This pin is connected to the micro USB port on your Pico, and taps into the USB 5 V power line before it's converted to 3.3 V to run your Pico's microprocessor. All three HC-SR501 pins should now be wired to your Pico: ground, signal, and power.

If you are using a different PIR, it might operate at a lower voltage. For example, many modules based on the AM312 PIR can be powered at 3.3 V, and you could power them from pin 36, 3V3(OUT), instead.

Programming your alarm

You'll need to program your Pico to recognise the sensor. Handily, this is no more difficult than reading a button — in fact, you can use the very same code. Start by creating a new program and importing the **machine** library so you can configure your Pico's GPIO pin, along with the **time** library which we'll use to set delays in the program:

```
import machine
import time
```

Then set up the pin you wired your HC-SR501 sensor to, GP28:

```
snsr_pir = machine.Pin(28, machine.Pin.IN, machine.Pin.PULL_DOWN)
```

Like the reaction game you made, a burglar alarm's inputs should act as an interrupt — stop the program doing whatever it was doing and react whenever the sensor is triggered. As before, start by defining a callback function to handle the interrupt:

```
def pir_handler(pin):
    time.sleep_ms(100)
    if pin.value():
        print("ALARM! Motion detected!")
```

Make sure you have correct indentation. The **time.sleep_ms(100)** and **if pin.value():** lines are used to prevent the alarm from being triggered by any jitter in the signal from the PIR sensor: if the pin is no longer active 100 milliseconds after the interrupt is triggered, it was probably a false positive. This process of smoothing out fluctuations is known as *debouncing*.

Finally, set up the interrupt itself. This isn't part of the handler function, so delete any spaces that Thonny inserts before it:

```
snsr_pir.irq(trigger=machine.Pin.IRQ_RISING, handler=pir_handler)
```

That's enough for now: interrupts stay active regardless of what the rest of the program is doing, so there's no need to add an infinite loop to keep your program running. Connect the Pico to your Raspberry Pi or computer, then click the **Run** icon and save the program to your Pico as **Burglar_Alarm.py**.

Wave your hand slowly over the PIR sensor: a message will print to the Shell area confirming that the sensor saw you. If you keep waving your hand, the message will keep printing — but with a delay between each time it's printed.

That delay isn't part of your program, but built into the HC-SR501: the sensor sends a trigger signal to your Pico's GPIO pin when it detects motion, and keeps the signal on for several seconds before dropping it. On most HC-SR501 sensors, you use a small screwdriver to turn one of its potentiometers to adjust the delay: turn it one way to decrease the delay and the other way to increase it. Check your sensor's documentation for which potentiometer to use.

Because your interrupt trigger is set to fire on the rising edge of the signal, the message is printed as soon as the PIR sensor sends its signal of 1 or 'high' (subject to the debouncing logic, that is). Even if more motion is detected, the interrupt won't fire again until the built-in delay has passed and the signal has returned to 0, or 'low'.

Printing a message to the Shell is enough to prove your sensor is working, but it doesn't make for much of an alarm. Real burglar alarms have lights and sirens that alert everyone around that something's wrong — and you can add the same to your own alarm.

Start by wiring an LED, of any colour, to your Pico as shown in **Figure 7-2** (disconnect the Pico from USB first). You need to connect the longer leg, the anode, to pin GP15 via a 330 Ω resistor — remember that without this resistor in place to limit the amount of current passing through the LED, you can damage both the LED and your Pico. The shorter leg, the cathode, needs to be wired to one of your Pico's ground pins — use your breadboard's ground rail and two male-to-male (M2M) jumper wires for this.

To set up the LED output, add a new line just below where you initialised the PIR sensor's pin:

```
led = machine.Pin(15, machine.Pin.OUT)
```

That's enough to configure the LED, but you'll need to make it light up. Add the following new line to your interrupt handler function, under the **print** line (it should be indented by eight spaces to match that):

```
for i in range(50):
```

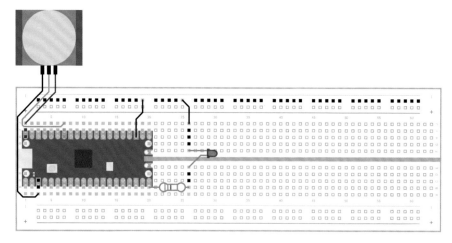

Figure 7-2   Adding an LED to the burglar alarm

You've just created a finite loop, one which will run 50 times. The letter **i** represents an *increment*, a value which goes up each time the loop runs, and which is populated by the instruction **range(50)**.

Give your new loop something to do, remembering that these lines underneath will need to be indented by a further four spaces (twelve in all), as they form both part of the loop you just opened and the interrupt handler function:

```
led.toggle()
time.sleep_ms(100)
```

The first of these lines is a feature of the **machine** library which lets you flip the value of an output pin, rather than set a value — so if the pin is currently 1 (high), toggling it will set it to 0 (low); if the pin is already 0, toggling it will set it to 1.

These two lines of code flash the LED on and off with a 100-millisecond — a tenth of a second — delay. The result is similar to the LED blinking program you wrote back in Chapter 4, *Physical computing with Raspberry Pi Pico*.

Your program will now look like this:

```
import machine
import time

snsr_pir = machine.Pin(28, machine.Pin.IN, machine.Pin.PULL_DOWN)
led = machine.Pin(15, machine.Pin.OUT)
```

```
def pir_handler(pin):
    time.sleep_ms(100)
    if pin.value():
        print("ALARM! Motion detected!")
        for i in range(50):
            led.toggle()
            time.sleep_ms(100)

snsr_pir.irq(trigger=machine.Pin.IRQ_RISING, handler=pir_handler)
```

> **VALUE VS TOGGLE**
>
> There are times when using the toggle function over setting a value makes sense, like when blinking an LED — but make sure you've thought through what you're trying to achieve first. If your project hinges on an output definitely being on or off at a given time — such as a warning light, or a pump which drains a water tank — always explicitly set the value rather than relying on a toggle.

Connect the Pico to USB, click the **Run** icon, then wave your hand over the PIR sensor again: you'll see the usual alarm message print to the Shell area, then the LED will begin rapidly flashing as a visual alert. Wait for the LED to stop flashing, then wave your hand over the PIR sensor again: the message will print again, and the LED will repeat its flashing pattern.

To make your burglar alarm even more of a deterrent, you can make it flash slowly even when there's no motion being detected — warning would-be intruders that your room is under observation. Go to the very bottom of your program and add in the following lines:

```
while True:
    led.toggle()
    time.sleep(5)
```

Click Run again, but leave the PIR sensor alone: you'll see the LED is now turning on for five seconds, then turning off for five seconds. This pattern will continue as long as the sensor isn't triggered; wave your hand over the PIR sensor and you'll see the LED rapidly flashing again, before going back to its slow-flash pattern.

This highlights a key difference between threads and interrupts: if you'd used threads, your program would still be trying to toggle the LED on and off with a five-second interval even as your PIR handler is flashing it on and off with a 100-millisecond interval. That's because threads run concurrently, side by side.

With interrupts, the main program is paused while the interrupt handler runs, so the five-second toggle code you've written stops until the handler has finished flashing the LED, then picks up from where it left off. Whether you need your code to pause or to keep running is the key to choosing between threads or interrupts, and will depend on exactly what your project is meant to do.

Inputs and outputs: putting it all together

Your burglar alarm now has a flashing LED to warn intruders away, and a way to see when it's been triggered without having to watch the Shell area for a message. Now all it needs is a siren — or, at least, a piezoelectric buzzer, which makes sound without deafening your neighbours.

Depending on which model you purchased, your piezoelectric buzzer will have either pins sticking out of the bottom or short wires attached to its sides. If the buzzer has pins, insert these into your breadboard so the buzzer is straddling the centre divide; if it has wires, place these in the breadboard and simply rest the buzzer on the breadboard.

If your buzzer's wires are long enough, try to connect them to the breadboard columns next to your Pico's pins; if not, use male-to-male (M2M) jumper wires to wire the buzzer as shown in **Figure 7-3**. Connect the red wire, or the positive pin marked with a + symbol, to pin GP14 at the bottom of your Pico, just to the left of the LED pin. Connect the black wire, or the negative pin marked with a minus (-) symbol (or the letters GND), to the ground rail of your breadboard.

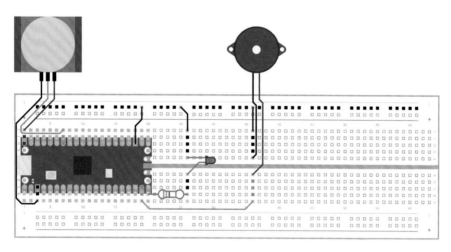

Figure 7-3 Wiring a two-wire piezoelectric buzzer

If your buzzer has three pins, connect the leg marked with a minus symbol (-) or the letters GND to the ground rail of your breadboard, the pin marked with

S or SIGNAL to pin GP14 on your Pico, and the remaining leg — which is usually the middle leg — to the 3V3 pin on your Pico.

If you run your program now, nothing will change: the buzzer will only make a sound when it receives power from your Pico's GPIO pins. Go back to the top of your program and initialise the buzzer just below where you initialised the LED:

```
buzzer = machine.Pin(14, machine.Pin.OUT)
```

Next, change your interrupt handler to add a new line below `led.toggle()` — remembering to indent it by twelve spaces to match:

```
buzzer.toggle()
```

Your program will now look like this:

```
import machine
import time

snsr_pir = machine.Pin(28, machine.Pin.IN, machine.Pin.PULL_DOWN)
led = machine.Pin(15, machine.Pin.OUT)
buzzer = machine.Pin(14, machine.Pin.OUT)

def pir_handler(pin):
    time.sleep_ms(100)
    if pin.value():
        print("ALARM! Motion detected!")
        for i in range(50):
            led.toggle()
            buzzer.toggle()
            time.sleep_ms(100)

snsr_pir.irq(trigger=machine.Pin.IRQ_RISING, handler=pir_handler)

while True:
    led.toggle()
    time.sleep(5)
```

Click Run and wave your hand over the PIR sensor: the LED will flash rapidly, as before, but this time it'll be accompanied by a beeping sound from the buzzer. Congratulations: that should be more than enough to scare an intruder away from ransacking your secret stash of sweets!

If you find your buzzer is clicking, rather than beeping, then you're using a *passive buzzer* rather than an *active buzzer*. An active buzzer has a component inside known as an *oscillator*, which rapidly moves the metal plate to make the buzzing sound; a passive buzzer lacks this component, meaning you need to replace it with some code of your own.

If you're using a passive buzzer, try this version of the program instead — it toggles the pin connected to the buzzer on and off very quickly, mimicking the effect of the oscillator in an active buzzer:

```
import machine
import time

snsr_pir = machine.Pin(28, machine.Pin.IN, machine.Pin.PULL_DOWN)
led = machine.Pin(15, machine.Pin.OUT)
buzzer = machine.Pin(14, machine.Pin.OUT)

def pir_handler(pin):
    time.sleep_ms(100)
    if pin.value():
        print("ALARM! Motion detected!")
        for i in range(50):
            led.toggle()
            for j in range(25):
                buzzer.toggle()
                time.sleep_ms(3)

snsr_pir.irq(trigger=machine.Pin.IRQ_RISING, handler=pir_handler)

while True:
    led.toggle()
    time.sleep(5)
```

Note that the new loop, which controls the buzzer, doesn't use the letter **i** to track the increment; that's because you're already using that letter for the outer loop — so it uses the letter **j**. The short delay of just three milliseconds, meanwhile, means the pin connected to the buzzer turns on and off rapidly enough for it to make a buzzing noise.

Try changing the delay to four milliseconds instead of three and you'll find the buzzer sounds at a lower pitch. Changing the delay changes the buzzer's oscillation frequency: a longer delay means it oscillates at a lower frequency making it a lower-pitched sound; a shorter delay makes it oscillate at a higher frequency, making it a higher-pitched sound.

Extending your alarm

Burglar alarms rarely cover a single room: instead, they use a network of multiple sensors to monitor multiple rooms from a single alarm system. Your Pico-based burglar alarm can work in exactly the same way, adding in multiple sensors to cover multiple areas at once.

You'll need a PIR sensor for each area you want to cover; in this example you'll add one more sensor for a total of two, but you can keep going and add as many sensors as you need.

Both of your sensors need 5 V power to work, but you've already used the VUSB pin on your Pico for the first sensor. If your breadboard has enough room, you could put a male-to-female (M2F) jumper wire next to the one connected to your first sensor and use it for the second; a neater approach, though, is to use the power rail on your breadboard.

Disconnect the first sensor's power wire from the breadboard end, and insert it into the power rail coloured red or marked with a plus (+) symbol. Take a male-to-male (M2M) jumper wire and connect the same power rail to your Pico's VBUS pin. Next, take a male-to-female (M2F) jumper wire and connect the power rail to your second PIR sensor's power input pin. If you're using a 3.3 V sensor, use 3V3(OUT) instead of VBUS.

Finally, wire up the ground pin and signal pin of your second PIR sensor as before — but this time connect the signal pin to pin GP18 on your Pico, as shown in **Figure 7-4** (breadboard column 17). Your circuit now has two sensors, each connected to a separate pin.

Setting your program up to read the second sensor as well as the first is as simple as adding two new lines. Start by initialising the second sensor, adding a new line below where you initialised the first sensor:

```
snsr_pir2 = machine.Pin(18, machine.Pin.IN, machine.Pin.PULL_DOWN)
```

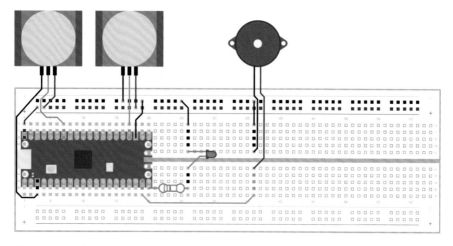

Figure 7-4   Adding a second PIR sensor to cover another room

Then create a new interrupt, again directly beneath your first interrupt (remember that you can have multiple interrupts with a single handler):

```
snsr_pir2.irq(trigger=machine.Pin.IRQ_RISING, handler=pir_handler)
```

Click **Run**, and wave your hand over the first PIR sensor: you'll see the alert message, the LED flash, and the buzzer sound as normal. Wait for them to finish, then wave your hand over the second PIR sensor: you'll see your burglar alarm respond in exactly the same way.

To make your alarm really smart, you can customise the message depending on which pin was responsible for the interrupt — and it works exactly the same way as in the two-player reaction game you wrote earlier. Go back to your interrupt handler and modify it so it looks like:

```
def pir_handler(pin):
    time.sleep_ms(100)
    if pin.value():
        if pin is snsr_pir:
            print("ALARM! Motion detected in bedroom!")
        elif pin is snsr_pir2:
            print("ALARM! Motion detected in living room!")
        for i in range(50):
            led.toggle()
            buzzer.toggle()
            time.sleep_ms(100)
```

Just as in the reaction game project in Chapter 6, *Reaction game*, this code uses the fact that an interrupt reports which triggered it: if the PIR sensor attached to pin GP28 is responsible, it will print one message; if it was the one on GP18, it will print another. Your finished program will look like this:

```python
import machine
import time

snsr_pir = machine.Pin(28, machine.Pin.IN, machine.Pin.PULL_DOWN)
snsr_pir2 = machine.Pin(18, machine.Pin.IN, machine.Pin.PULL_DOWN)
led = machine.Pin(15, machine.Pin.OUT)
buzzer = machine.Pin(14, machine.Pin.OUT)

def pir_handler(pin):
    time.sleep_ms(100)
    if pin.value():
        if pin is snsr_pir:
            print("ALARM! Motion detected in bedroom!")
        elif pin is snsr_pir2:
            print("ALARM! Motion detected in living room!")
        for i in range(50):
            led.toggle()
            buzzer.toggle()
            time.sleep_ms(100)

snsr_pir.irq(trigger=machine.Pin.IRQ_RISING, handler=pir_handler)
snsr_pir2.irq(trigger=machine.Pin.IRQ_RISING, handler=pir_handler)

while True:
    led.toggle()
    time.sleep(5)
```

If you're using a passive, rather than active, buzzer, remember you'll need to change the buzzer toggle to a loop to have it beep. Click **Run** and wave your hand over one sensor, then the other, to see both messages print to the Shell area. Congratulations: you now know how to build a modular burglar alarm capable of covering as many areas as you need!

CHALLENGE: CUSTOMISATION

Can you extend the burglar alarm with another PIR sensor? What about adding another LED, or another buzzer? Can you change the messages that print to match the areas you're covering with each sensor? Can you make the buzzer sound for longer, or for less time? Can you think of any other sensors, apart from a PIR sensor, that might work well in a burglar alarm?

Chapter 8

Temperature gauge

Use your Raspberry Pi Pico's built-in ADC to convert analogue inputs, and also to read its internal temperature sensor

In previous chapters you've been using the digital inputs on your Raspberry Pi Pico. A digital input is either on or off, a *binary* state. When a push-button switch is pressed, it changes a pin from low, off, to high, on; when a passive infrared sensor detects motion, it does the same.

Your Pico can accept another type of input signal, though: *analogue input*. Whereas digital is only ever either on or off, an analogue signal can be anything from completely off to completely on — a range of possible values. Analogue inputs are used for everything from volume controls to gas, humidity, and temperature sensors — and they work through a piece of hardware known as an *analogue-to-digital converter* (*ADC*).

In this chapter you'll learn how to use the ADC on your Pico — and how to tap into its internal temperature sensor to build a data-logging heat-measurement gadget. You'll also learn a technique for creating an analogue-like output. For this you'll need your Pico; an LED of any colour and 330 Ω resistor; a 10 kΩ potentiometer; and a selection of male-to-male (M2M) jumper wires. You'll also need a micro USB cable to connect your Pico to your Raspberry Pi or other computer running Thonny.

The analogue-to-digital converter

Raspberry Pi Pico's RP2040 microcontroller is a digital device, like all micro-controllers: it is built up of thousands of *transistors*, tiny switch-like devices which are either on or off. As a result, there's no way for your Pico to truly understand an analogue signal — one which can be anything on a spectrum between fully off and fully on — without relying on an additional piece of hardware: the analogue-to-digital converter (ADC).

As the name suggests, an analogue-to-digital converter takes an analogue signal and changes it to a digital one. You won't see the ADC on your Pico, no matter how closely you look: it's built into RP2040 itself. Many microcontrollers have their own ADCs, just like RP2040, and the ones that don't, can use an external ADC connected to one or more of their digital inputs.

An ADC has two key features: its *resolution*, measured in digital bits, and its *channels*, or how many analogue signals it can accept and convert at once. The ADC in your Pico has a resolution of 12 bits, meaning that it can transform an analogue signal into a digital signal as a number ranging from 0 to 4095 — though MicroPython transforms this to a 16-bit number ranging from 0 to 65,535, so that it returns the same range of values as the ADC on other MicroPython microcontrollers. It has three channels brought out to the GPIO pins: GP26, GP27, and GP28, which are also known as GP26_ADC0, GP27_ADC1, and GP28_ADC2 for analogue channels 0, 1, and 2. There's also a fourth ADC channel, which is connected to a temperature sensor built into RP2040; you'll find out more about that later in the chapter.

WHY 65,535?

The number 65,535 looks strange at first glance — why that, and why not simply 0–100? The answer ties into the fact your Pico works on a binary number system, where the only possible values for a digit are 0 or 1. A 16-bit binary number is made up of 16 digits, and the maximum possible value is 16 ones: **1111111111111111**. If you convert that into decimal numbers, the 0–9 counting system humans use, you get 65,535.

Reading a potentiometer

Every pin connected to your Pico's analogue-to-digital converter can also be used as a simple digital input or output; to use it as an analogue input, you'll need an analogue signal — and you can easily make one with a potentiometer.

There are various types of potentiometer available: some, like the ones in the HC-SR501 passive infrared sensor you used in Chapter 7, *Burglar alarm*, are designed to be adjusted with a screwdriver; others, often used for volume controls and other inputs, have knobs or sliders. The most common type has a small, usually plastic, knob coming out of the top or front: this is known as a *rotary potentiometer*.

Pick up your potentiometer and turn it over: you'll see it has three pins which fit in the breadboard. Depending on how you connect these pins, the potentiometer responds in two different ways. Unplug your Pico from USB, then insert the potentiometer into your breadboard, being careful not to bend the pins. Wire the middle pin to pin GP26_ADC0 on your Pico using a male-to-male (M2M) jumper wire (see **Figure 8-1**) If your breadboard is oriented as shown, it'll be above the Pico in column ten. Finally, take two more jumper wires and wire either of the potentiometer's outer pins to your breadboard's power rail and the power rail to your Pico's 3V3 pin.

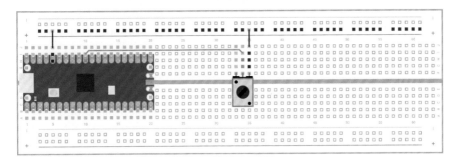

Figure 8-1   A potentiometer wired with two pins connected

Open Thonny and begin a new program:

```
import machine
import time
```

Like the digital general-purpose input/output (GPIO) pins, the analogue input pins are handled by the **machine** library — and just like the digital pins, they need to be set up before you can use them. Continue your program:

```
potentiometer = machine.ADC(26)
```

This configures pin GP26_ADC0 as the first channel, ADC0, on the analogue-to-digital converter. To read from the pin, set up a loop:

```
while True:
    print(potentiometer.read_u16())
    time.sleep(2)
```

In this loop, reading the value of the pin and printing it take place on a single line: this is a more compact alternative to reading the value into a variable and then printing the variable, but only works if you don't want to do anything with the reading other than print it — which is exactly what this program needs at the moment.

Reading an analogue input is just like reading a digital input, except for one thing: when you read a digital input you use **read()**, but you're reading this analogue input with **read_u16()**. That last part, **u16**, simply warns you that rather than receiving a binary 0 or 1 result, you'll receive an *unsigned 16-bit integer* — a whole number between 0 and 65,535.

Connect your Pico to your computer or Raspberry Pi over USB, then save your program as **Potentiometer.py** and click the **Run** icon. Watch the Shell: you'll see your program print out a large number, likely over 60,000. Try turning the potentiometer all the way in one direction: depending on the direction you turned the knob and which outer leg you connected, the number will go up or down. Turn it the other way: the value will change in the opposite direction.

No matter which way you turn it, though, it will never get anywhere near 0. That's because with only two legs connected, the potentiometer is acting as a component known as a *variable resistor* or *varistor*. A varistor is a resistor with a value you can change — in the case of a 10 kΩ potentiometer, between 0 Ω and 10,000 Ω. The higher the resistance, the less voltage from the 3V3 pin reaches your analogue input — so the number goes down. The lower the resistance, the more voltage reaches your analogue input — so the number goes up.

A potentiometer works by having a conductive strip inside, connected to the two outer pins, and a *wiper* or *brush* connected to the inner pin (**Figure 8-2**). As you turn the knob, the wiper moves closer to one end of the strip and further away from the other. The further the wiper gets from the end of the strip you wired to your Pico's 3V3 pin, the higher the resistance; the closer it gets, the lower the resistance.

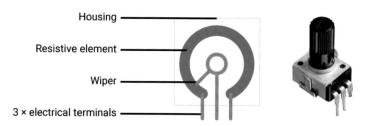

Housing

Resistive element

Wiper

3 × electrical terminals

Figure 8-2   How a potentiometer works

Varistors are extremely useful components, but there's a drawback: you'll notice no matter how far you turn the knob in either direction, you can never get a value of 0 — or anywhere close to it. That's because a 10 kΩ resistor isn't strong enough to drop the 3V3 pin's output to 0 V. You could look for a bigger potentiometer with a higher maximum resistance, or you could simply wire your existing potentiometer up as a *voltage divider*.

A potentiometer as a voltage divider

The unused pin on your potentiometer isn't there for show: adding a connection to that pin to your circuit completely changes how the potentiometer works. Click the Stop icon to stop your program, disconnect your Pico from USB, and grab two male-to-male (M2M) jumper wires. Use one to connect the unused pin of your potentiometer to your breadboard's ground rail as shown in **Figure 8-3**. Take the other and connect the ground rail to a GND pin on the Pico, such as the one in column 3.

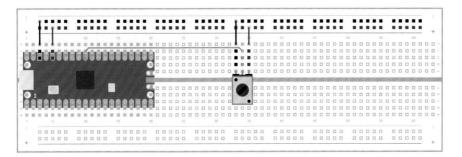

Figure 8-3   Wiring the potentiometer as a voltage divider

Connect your Pico to USB again, then click the **Run** icon to restart your program. Turn the potentiometer knob again, all the way one direction then all the way the other. Watch the values that are printed to the Shell area: unlike before, they're now going from near-zero to nearly a full 65,535 — but why?

Adding the ground connection to the other end of the potentiometer's conductive strip has created a voltage divider: previously, the potentiometer was simply acting as a resistor between the 3V3 pin and the analogue input pin, it's now dividing the voltage between the 3.3 V output by the 3V3 pin and the 0 V of the GND pin. Turn the knob fully one direction, you'll get 100 percent of the 3.3V; turn it fully the other way, 0 percent.

ZERO'S THE HARDEST NUMBER

If you can't get your Pico's analogue input to read exactly zero or exactly 65,535, don't worry — you haven't done anything wrong! All electronic components are built with a *tolerance*, which means any claimed value isn't going to be precise. In the case of the potentiometer, it will likely never reach exactly 0 or 100 percent of its input — but it will get you very close!

The number you see printed to the Shell is a decimal representation of the raw output of the analogue-to-digital converter — but it's not the friendliest way to see it, especially if you forget that 65,535 means 'full voltage'.

There's an easy way to fix that, though: a simple mathematical equation. Go back to your program, and add the following above your loop:

```
conversion_factor = 3.3 / (65535)
```

This sets up a mathematical way to convert the number that the analogue-to-digital converter gives you into a fair approximation of the actual voltage it represents. The first number is the maximum possible voltage that the pin can expect: 3.3 V, from your Pico's 3V3 pin; the second number is the maximum value the analogue input reading can be, 65,535.

Taken all together, the conversion factor is a number created by '3.3 divided by 65,535' — the maximum possible voltage divided by the range of values the analogue-to-digital converter reports, which is in turn a feature of its resolution in bits.

With your conversion factor set up, you simply need to use it in your program. Go back to your loop, and edit it to read:

```
while True:
    voltage = potentiometer.read_u16() * conversion_factor
    print(voltage)
    time.sleep(2)
```

The first line inside the loop takes a reading from the potentiometer via the analogue input pin, and multiplies it — the * symbol — by the conversion factor you set up earlier in the program, storing the result as the variable voltage. That variable is then printed to the Shell, in place of the raw reading you used earlier.

Your finished program will look like this:

```
import machine
import time

potentiometer = machine.ADC(26)
conversion_factor = 3.3 / (65535)

while True:
    voltage = potentiometer.read_u16() * conversion_factor
    print(voltage)
    time.sleep(2)
```

Click the **Run** icon. Turn the potentiometer all the way in one direction, then the other. Watch the numbers being printed to the Shell area: you'll see that when the potentiometer is all the way one way, the numbers get very close to zero; when it's all the way the other way, they get very close to 3.3. These numbers represent the actual voltage being read by the pin — and as you turn the knob of the potentiometer, you're dividing the voltage smoothly between minimum and maximum, 0 V to 3.3 V.

Congratulations: you now know how to wire a potentiometer as both a varistor and a voltage divider, and how to read analogue inputs as both a raw value and a voltage!

Measuring temperatures

Your Raspberry Pi Pico's RP2040 microcontroller has an internal temperature sensor, which is read on the fourth analogue-to-digital converter channel. Like the potentiometer, the output of the sensor is a variable voltage: as the temperature changes, so does the voltage.

Start a new program, and import the machine and time libraries:

```
import machine
import time
```

Set up the analogue-to-digital converter again, but this time rather than using the number of a pin, use the channel number of the ADC associated with the internal temperature sensor:

```
sensor_temp = machine.ADC(4)
```

You'll need your conversion factor again, to change the raw reading from the sensor into a voltage value, so add that:

```
conversion_factor = 3.3 / (65535)
```

Then set up a loop to take readings from the analogue input, apply the conversion factor, and store them in a variable:

```
while True:
    reading = sensor_temp.read_u16() * conversion_factor
```

Rather than print the reading directly, though, you need to do a second conversion — to take the voltage reported by the analogue-to-digital converter and convert it into degrees Celsius:

```
temperature = 27 - (reading - 0.706)/0.001721
```

This is another mathematical equation, and one which is specific to the temperature sensor in RP2040. The values are taken from a technical document called a data sheet or data book: all electronic components have a data sheet, which is normally available on request from the manufacturer. You can view RP2040's data sheet in the Pico documentation at **rptl.io/rp2040-get-started** It's packed full of information on how the microcontroller works, though it's aimed at engineers, so it is deeply technical.

Finally, finish your loop:

```
    print(temperature)
    time.sleep(2)
```

Your program will now look like this:

```
import machine
import time

sensor_temp = machine.ADC(4)
conversion_factor = 3.3 / (65535)

while True:
    reading = sensor_temp.read_u16() * conversion_factor
    temperature = 27 - (reading - 0.706)/0.001721
    print(temperature)
    time.sleep(2)
```

save your program as **Temperature.py** and click the **Run** icon. Watch the Shell area: you'll see numbers being printed which represent the temperature reported by the sensor in degrees Celsius.

Try gently pressing the tip of your finger to RP2040, the largest black chip in the middle of your Pico, and holding it there: the warmth of your finger should make the chip warmer, and the temperature will rise. Remove your finger from the chip, and the temperature will fall again.

Congratulations — you've turned your Pico into a thermometer!

Fading an LED with PWM

The analogue-to-digital converter in your Pico only works one way: it takes an analogue signal and converts it to a digital signal the microcontroller can understand. If you want to go the other way, and have your digital microcontroller create an analogue output, you'd normally need a digital-to-analogue converter (DAC) — but there's a way to 'fake' an analogue signal, using a feature called *pulse-width modulation* or *PWM*.

A microcontroller's digital output can only ever be on or off, 0 or 1. Turning a digital output on and off is known as a *pulse* and by altering how quickly the pin turns on and off you can change, or *modulate*, the *width* of these pulses — hence 'pulse-width modulation'.

Every GPIO pin on your Pico is capable of pulse-width modulation, but the microcontroller's pulse-width modulation block is made up of eight slices, each with two outputs. Look at **Figure 8-4**: you'll see that each pin has a letter and a number. The number represents the PWM slice connected to that pin; the letter represents which output of the slice is used.

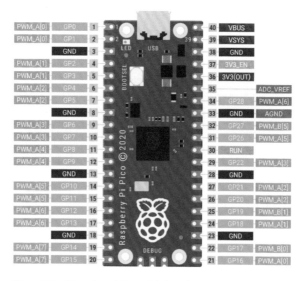

Figure 8-4   The pulse-width modulation pins

If that sounds confusing, don't worry: all it means is that you need to make sure you keep track of the PWM slices and outputs you're using, making sure to only connect to pins with a letter and number combination you haven't already used. If you're using PWM_A[0] on pin GP0 and PWM_B[0] on pin GP1, things will work fine, and will continue to work if you add PWM_A[1] on pin GP2; if you try to use the PWM channel on pin GP0 and pin GP16, though, you'd run into problems as they're both connected to PWM_A[0].

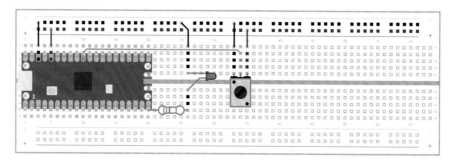

Figure 8-5   Adding an LED

With your Pico disconnected from USB, take an LED of any colour and a 330 Ω current-limiting resistor, and put them in the breadboard as shown in **Figure 8-5**. Wire the longer leg of the LED, the anode, to pin GP15 via the 330 Ω resistor, and wire the shorter leg to the ground pin of your Pico. Now you can plug your Pico back into USB.

Go back to your first program by clicking on its tab just under Thonny's toolbar; if you'd already closed it, click the **Open** icon and load **Potentiometer.py** from your Pico. Delete the line that starts with `conversion_factor =`, and replace it with this:

```
led = machine.PWM(machine.Pin(15))
```

This creates an LED object on pin GP15, but with a difference: it activates the pulse-width modulation output on the pin, channel B[7] — the second output of the eighth slice (slices are counted starting from zero).

You'll also need to set the frequency, one of the two values you can change to control, or modulate, the pulse width. Add the following line immediately below the previous line:

```
led.freq(1000)
```

This sets a frequency of 1000 hertz — one thousand cycles per second. Next, go to the bottom of your program and delete the line starting with `voltage =` and the `print(voltage)` line before adding the following. Remember to keep it indented by four spaces so it forms part of the nested code within the loop:

```
    led.duty_u16(potentiometer.read_u16())
```

Next, replace the **2** in `time.sleep(2)` with `0.1`:

```
    time.sleep(0.1)
```

This line takes a raw reading from the analogue input connected to your potentiometer, then uses it as the second aspect of pulse-width modulation: the *duty cycle*. The duty cycle controls the pin's output: a 0 percent duty cycle leaves the pin switched off for all 1000 pulses per second, and effectively turns the pin off; a 100 percent duty cycle leaves the pin switched on for all 1000 pulses per second, and is functionally equivalent to just turning the pin on as a fixed digital output; a 50 percent duty cycle has the pin on for half the pulses and off for half the pulses.

Click **Run** and watch the LED as you turn the potentiometer: the LED will grow brighter with the potentiometer turned all the way one way, and gradually dimmer as you turn it the other. That's because the reading taken from the analogue pin connected to the potentiometer is being turned into a value for the PWM signal's duty cycle: a low duty cycle is like a low voltage on an analogue output, making the LED dim; a high duty cycle is like a high voltage, making the LED bright.

To make it so you can properly control the LED's brightness, you need to map the value from the analogue input to a range the PWM slice can understand. The best way to do this is to tell MicroPython that you're passing the duty cycle value as an unsigned 16-bit integer, the same number format as you receive from your Pico's analogue input pin. This is why the previous line of code used **duty_u16()** instead of **duty()**.

Your finished program will look like this:

```python
import machine
import time

potentiometer = machine.ADC(26)
led = machine.PWM(machine.Pin(15))
led.freq(1000)

while True:
    led.duty_u16(potentiometer.read_u16())
    time.sleep(0.1)
```

Click the **Run** icon and try turning the potentiometer all the way one way, then all the way the other. Watch the LED: this time, unless you're using a logarithmic potentiometer, you'll see the LED's brightness change smoothly from completely off at one end of the potentiometer knob's limit to fully lit at the other.

Congratulations: you've not only mastered analogue inputs, but you can now create the equivalent to an analogue output using pulse-width modulation!

CHALLENGE: CUSTOMISATION

Can you combine your two programs, and have the LED's brightness controlled by the temperature reading from the on-board temperature sensor? Can you remember how many analogue inputs your Pico has? What about PWM outputs? Try adding another analogue sensor to your Pico — something like a *light-dependent resistor (LDR)*, *gas sensor*, or *barometer* — and have your program read that instead of the potentiometer.

Chapter 9

Data logger

Untether Raspberry Pi Pico from the computer to make it a fully portable temperature-logging device

Throughout this book, you've been using your Raspberry Pi Pico connected to your Raspberry Pi or other computer via its micro USB port. As with all microcontrollers, though, there's no reason your Pico must be tethered in this way: it's a fully functional self-contained system, with processing capabilities, memory, and everything it needs to work on its own.

In this chapter you'll learn how to use the file system to create, write to, and read from files, allowing you to put your Pico anywhere you like and record data for later access — turning it into what is known as a *data logger*. For this you'll only need your Pico and, if you want to use it away from your Raspberry Pi, a micro USB charger or battery pack; once you have finished the chapter, you can connect additional sensors if you want to expand your project.

The file system

The file system is where your Pico stores all the programs you've been writing. It's equivalent in function to the microSD card in your Raspberry Pi, or the hard drive or solid-state drive in your laptop or desktop computer: it's a form of *non-volatile* storage, which means that whatever you save there stays in place even when you unplug your Pico's micro USB cable.

Connect your Pico to your Raspberry Pi (or computer) and load Thonny if it's not open. Click **Open** then click **Raspberry Pi Pico** in the pop-up that appears. You'll see a list of all the programs you've written so far, stored on your Pico's file system. You're not going to open one right now, so click **Cancel**.

Click at the bottom of the Shell area to start working with your Pico in interactive mode. Type:

```
file = open("test.txt", "w")
```

This tells MicroPython to open a file called **test.txt** for writing — the **"w"** part of the instruction. You won't see anything print to the Shell area when you press **ENTER** at the end of the line, because although you've opened the file, you haven't done anything with it yet. Type:

```
file.write("Hello, File!")
```

When you press **ENTER** at the end of this line, you'll see the number **12** appear (**Figure 9-1**). That's MicroPython confirming to you that it has written twelve bytes to the file you opened. Count the number of characters in the message you wrote: including the letters, comma, space, and exclamation mark, there are twelve — each of which takes up a single byte.

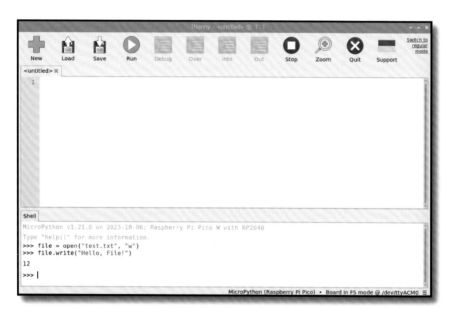

Figure 9-1   The size of the data you've written is printed to the Shell area

When you're done writing to a file, you need to close it — this ensures that the data you've told MicroPython to write is actually written to the file system. If you don't close the file, the data might not have been written yet — a bit like writing a letter in LibreOffice Writer or another word processor and forgetting to save it. Type:

```
file.close()
```

Your file is now safely stored on your Pico's file system. Click the **Open** icon on Thonny's toolbar, click **Raspberry Pi Pico**, and scroll through the list of files until you find **test.txt**. Click on it, then click **OK** to open it: you'll see your message pop up in a new Thonny tab.

You don't have to use **Open** to read files, though: you can open the file right in MicroPython itself. Click back into the bottom of the Shell area and type:

```
file = open("test.txt")
```

You'll notice that this time around there's no **"w"**: that's because instead of writing to the file, you're going to be reading it. You could replace the **"w"** with an **"r"**, but MicroPython defaults to opening a file in read mode — so it's fine to simply leave that part of the instruction off. Next, type:

```
file.read()
```

You'll see the message you wrote to the file print to the Shell area (**Figure 9-2**). Congratulations: you can read and write files on your Pico's file system!

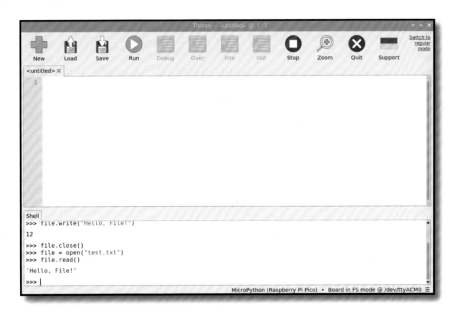

Figure 9-2   Printing the stored message to the Shell area

Before you finish, close the file — it's not as important to properly close a file after reading it as it is when writing, but it's a good habit to get into:

```
file.close()
```

Logging temperatures

Now you know how to open, write to, and read from files, you have everything you need to build a data logger on your Pico. Click the **New** icon to start a new program in Thonny, and start your program by typing:

```
import machine
import time

sensor_temp = machine.ADC(machine.ADC.CORE_TEMP)

conversion_factor = 3.3 / (65535)
reading = sensor_temp.read_u16() * conversion_factor
temperature = 27 - (reading - 0.706)/0.001721
```

You might recognise this code: it's the same as you used in Chapter 8, *Temperature gauge* to read from your Pico's on-board temperature sensor. The readings from the sensor are the data you're going to be logging to the file system, so you don't want to simply print them out as you did before.

Start by opening a file for writing by adding the following line at the bottom:

```
file = open("temps.txt", "w")
```

If the file doesn't already exist on the file system, this creates it; if it does, it overwrites it — emptying its contents ready for you to write new data.

WARNING

Opening a file for writing in MicroPython will delete anything you've already stored in it. Always make sure you've opened the file for reading and saved the contents somewhere if you want to keep it!

Now you need to write something to the file — the value you got from the temperature sensor:

```
file.write(str(temperature))
```

Rather than writing a fixed string in quotes, as you did before, this time you're converting the variable **temperature** — which is a floating-point number (a number with a decimal point in it) — to a string, then writing that to the file.

As before, to make sure the data is written, you need to close the file:

```
file.close()
```

Click **Run**, then save your program to the Raspberry Pi Pico as **Datalogger.py**. The program will only take a few seconds to run; when the **>>>** prompt reappears at the bottom of the Shell area, click into it and type the following to open and read your new file:

```
file = open("temps.txt")
file.read()
file.close()
```

You'll see the temperature reading your program took appear in the Shell (**Figure 9-3**). Congratulations: your data logger works!

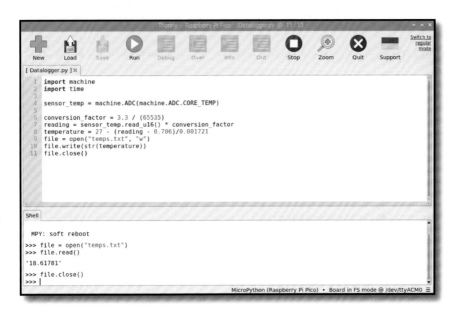

Figure 9-3   Your file records the temperature at the time the measurement was taken

A data logger that only logs a single reading — a *datum* — isn't that useful, though. To make your data logger more powerful, you need to modify it so it takes lots of readings. Click **Run** again, and read the file again:

```
file = open("temps.txt")
file.read()
file.close()
```

Notice how there's still only one reading in the file. When your program opened the file for writing again, it automatically wiped its previous contents — meaning that each time your program runs, it will wipe the file and store a single reading.

To fix that, you need to modify your program. Start by clicking and dragging your mouse cursor to highlight the lines:

```
reading = sensor_temp.read_u16() * conversion_factor
temperature = 27 - (reading - 0.706)/0.001721
```

When you've highlighted both lines completely, let go of the mouse button and type **CTRL+X** (or **COMMAND+X** on Mac) to cut the lines; you'll see them disappear. Go to the bottom of your program and delete everything after:

```
file = open("temps.txt", "w")
```

Now type:

```
while True:
```

After pressing **ENTER** at the end of that line, paste the two lines you cut earlier (**CTRL-V** or **COMMAND-V**). You'll see them appear, which saves you having to type them in — but only the first line will be indented correctly under the infinite loop you just created. Put your cursor at the start of the second line, then press **SPACE** four times to indent the line properly. Move your cursor to the end of the line and press **ENTER**. Type the following line, making sure it's properly indented:

```
file.write(str(temperature))
```

Now, though, you're going to need to do something new. If you close the file as you did before, you won't be able to write to it again without reopening it and wiping its contents. If you don't close the file, the data will never actually get written to the file system.

The solution: flush the file, rather than close it. Type:

```
file.flush()
```

When you're writing to a file but the data isn't actually being written to the file system, it's stored in what's known as a *buffer* — a temporary storage area. When you close the file, the buffer is written to the file in a process known as *flushing*. Using `file.flush()` is equivalent to `file.close()`, in that it flushes the contents of the buffer into the file — but unlike `file.close()`, the file remains open for you to write more data to it later.

Now you just need to pause your program between readings:

```
time.sleep(10)
```

Your finished program will look like this:

```python
import machine
import time

sensor_temp = machine.ADC(machine.ADC.CORE_TEMP)

conversion_factor = 3.3 / (65535)
file = open("temps.txt", "w")
while True:
    reading = sensor_temp.read_u16() * conversion_factor
    temperature = 27 - (reading - 0.706)/0.001721
    file.write(str(temperature))
    file.flush()
    time.sleep(10)
```

Click **Run**, count to 60, then click **Stop**. Run this code in the Shell area:

```python
file = open("temps.txt")
file.read()
file.close()
```

The good news is that your program worked, and you've logged multiple readings — around six, depending on how fast you counted. The bad news is that they're all mushed together (**Figure 9-4**) — making them difficult to read.

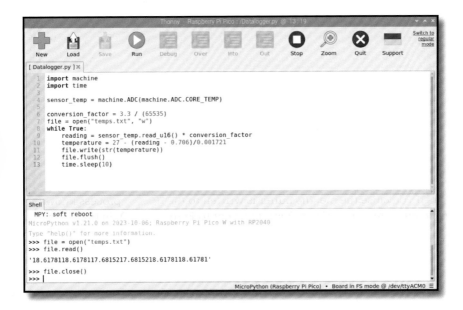

Figure 9-4   All the measurements are there, but the formatting makes it difficult to read

To fix that problem, you need to format the data as it's written to the file. Go back to the **file.write()** line in your program, and modify it so it looks like:

```
file.write(str(temperature) + "\n")
```

The plus symbol (**+**) tells MicroPython that you want to append what follows, concatenating the two strings together; **"\n"** is a special string known as a *control character* — it acts as the equivalent of pressing the **ENTER** key, meaning that each line in your data log should be on its own separate line.

Click the **Run** icon, count to 60 again, and click **Stop**. Open and read your file:

```
file = open("temps.txt")
file.read()
file.close()
```

You've made progress, but it's still not right: the **\n** control character isn't acting like a press of **ENTER**, but printing as two visible characters (**Figure 9-5**). That's because **file.read()** is bringing in the raw contents of the file, and making no attempt at formatting it for the screen.

Figure 9-5   You're getting closer to an easy-to-read printout here

To fix the formatting problem, wrap **file.read** in a **print()** function:

```
file = open("temps.txt")
```

```
print(file.read())
file.close()
```

This time you'll see each reading print out on its own line, neatly formatted and easy to read (**Figure 9-6**).

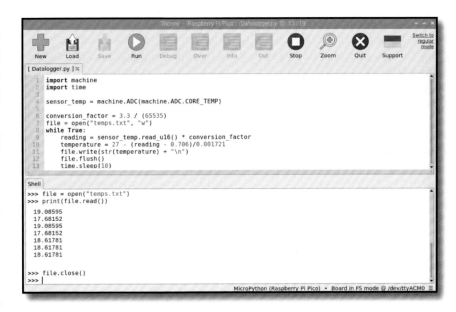

Figure 9-6   Now you can read all the temperatures your data logger has captured

Congratulations: you've built a data logger which can take multiple readings and store them on your Pico's file system!

FILE STORAGE

Your Pico's file system is 1.375MiB in size, meaning it can hold 1,441,792 bytes of data. Every file you save on your Pico, including the data logger's storage file, takes up room. How long it takes to fill the storage will depend on how many other files you have and how often your data logger saves a reading: at nine bytes per reading every ten seconds, you'll fill 1.375MiB in around 18.5 days; if you took a reading every minute, your data logger could run for around 111 days; once an hour, and your data logger could run more than 18 years!

If you'd like your data logger to write to the same file every time you run the program, change the **"w"** in `file = open("temps.txt", "w")` to an **"a"**. This creates the file if it doesn't exist, but appends to it if it does.

Running Headless

Your Pico's file system works regardless of whether or not it's connected to your Raspberry Pi or another computer. If you have a micro USB mains charger or a USB battery pack with a micro USB cable, you can take your data logger to any room in your house and have it run by itself — but you'll need a way to get your program running without having to click the Run icon in Thonny.

For use without a connected computer — known as *headless operation* — you can save your program under a special file name: **main.py**. When MicroPython finds a file called **main.py** (**boot.py** is similar, but it runs before the Pico is fully-configured) in its file system, it runs that automatically every time it's powered on or reset — without you having to click Run.

In Thonny, after stopping the program if running, click the **File** menu then **Save As**. Click **Raspberry Pi Pico** in the pop-up that appears, then type **main.py** as the file name before clicking **Save**. At first, nothing will seem to happen: that's because Thonny puts your Pico into interactive mode, which stops it from automatically running the program you just saved.

To force the program to run, click into the bottom of the Shell area and press **CTRL+D**. This sends your Pico a *soft reset* command, which will break it out of interactive mode and start the program running. Find something else to do for five minutes or so, then press the **Stop** icon and open your data log:

```
file = open("temps.txt")
print(file.read())
file.close()
```

You'll see a list of temperature readings, even though you didn't click the **Run** icon — because your program ran automatically when your Pico reset.

If you have a micro USB charger or USB battery pack, disconnect your Pico from your Raspberry Pi, take it to another room, and connect it to the charger or battery pack. Leave it there for ten minutes, then come back and unplug it. Take it back to your Raspberry Pi, plug it back in, and read your file again: you'll see the readings from the other room, proving that your Pico can run perfectly well without your Raspberry Pi helping it along.

Congratulations: your data logger is now fully functional and wholly portable, ready to go with you wherever you need to record data!

CHALLENGE

Can you change your program to record data from an external sensor connected to one of your Pico's ADC pins? Can you have your program write a title at the start of the file, so it's easier to see what the values mean? Can you write a program which logs how many times a push-button switch has been pressed? Can you figure out a way, such as copy-and-paste, to get your data into LibreOffice Calc or another spreadsheet program to create a chart?

Chapter 10

Digital communication protocols: I2C and SPI

Explore these two popular communication protocols and use them to display data on an OLED display

So far we've looked at how to work with a few common bits of hardware, but as you build more projects on your own, you'll probably want to branch out to use all sorts of different sensors, actuators, and displays. How will you communicate with these? Even if there's a MicroPython library you can use that converts the low-level functions into an easy-to-use package, some of the low-level interfaces require a bit more thought to work with.

There are a couple of standard low-level interfaces for connecting digital devices available in MicroPython: Inter-Integrated Circuit (I2C) and Serial Peripheral Interface (SPI). In many ways, they're very similar in that they both define a way of establishing a two-way interface between two devices. In fact, many parts come in versions with either interface, so you can pick the one that's right for your project. In both cases, there's one device that controls the communication (your Pico) and one (or more) that waits for instructions from the main device. However, there are a few differences, which you'll learn about as you use them.

> **VOLTAGE LEVELS**
>
> Your Pico's GPIO pins work at 3.3 volts. Applying a higher voltage to them may damage them. Fortunately, this is a common voltage to work at and a large proportion of the devices you come across will work at 3.3 volts. However, before plugging some new hardware into your Pico, always double-check that it's a 3.3 V device, as both I2C and SPI devices can run at 5 V sometimes and this will damage your Pico.

I2C

Communication over I2C takes place on two wires: a clock (marked as SCL or SCK) and a data channel (usually marked SDA).

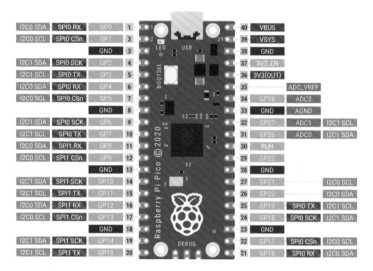

Figure 10-1   The pinout for the Pico. I2C functions are shown in light blue; SPI in pink

I2C will only work with certain pins on the Pico. There are a few choices; look at the pinout diagram for the options (**Figure 10-1**). There are two I2C buses (I2C0 and I2C1), and you can use either or both. In our example, we'll use I2C0 — with GP0 for SDA, and GP1 for SCL.

To demonstrate the protocols, we'll use an affordable *organic light-emitting diode* (*OLED*) module based on the SSD1306 controller, which are available from a wide variety of suppliers in both I2C and SPI.

You will sometimes find a module that has both interfaces on the same chip. you can select which interface to use with either a jumper or a small drop of solder that acts as a bridge across two pads. Remember when we warned you in "Soldering the headers" on page 6 to avoid bridges? This is one of those rare cases one might be useful!

The first example assumes you're using a 0.91" 128x32 pixel SSD1306 display. This model typically has four pins: 3.3 V (labelled VCC), ground (GND), and two pins for I2C (SDA and SCL). You can display text or graphics on it.

Wiring I2C is just a case of connecting the SDA pin on the Pico with the SDA pin on the OLED and the SCL pin to the OLED's SCK. Because of the way I2C handles communication, there also needs to be a resistor connecting SDA to

3.3 V and SCL to 3.3 V. Typically these are about 4.7 kΩ. However, with our device, these resistors are already included, so we don't need to add any extra ones. Disconnect your Pico from USB while wiring this up, but plug it back in when you're ready to program.

You'll need to install a library that isn't included with MicroPython to work with this module. Launch Thonny if it isn't already running. If Thonny is in simple mode, click the **Switch to regular mode** link in the upper right of the Thonny window. This reveals many advanced features, including the ability to install packages. Click the **Tools** menu at the top of the screen or Thonny window, then select **Manage packages**. In the window that appears, type **ssd1306**, and then press the **Search** button. Click the underlined name of the library in the search results, then click **Install** to download the library to your Pico's **/lib** directory. You'll need to do this for each Pico that you want to use with this module. Click **Close** to dismiss the package manager.

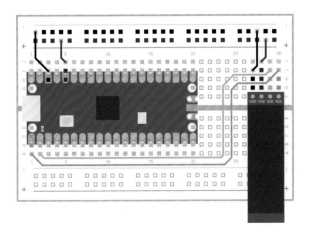

Figure 10-2   Wiring up a 0.91" SSD1306 module for I2C

With this wired up (see **Figure 10-2**), displaying information on the screen is as simple as running this code:

```
import machine
import ssd1306

sda = machine.Pin(0)
scl = machine.Pin(1)
i2c = machine.I2C(0, sda=sda, scl=scl, freq=400000)
display = ssd1306.SSD1306_I2C(128, 32, i2c)

display.text("Hello, Pico!", 0, 0, 1)
display.show()
```

This code configures the I2C connection, and creates an object (**display**) to represent the 128x32 pixel device. Next, it sends some text to the display, and tells the display to show it. The three arguments to the text function are the X and Y position of the text followed by the number of the colour to use. Your SSD1306 module is monochrome, so there's only two colours you can choose (0 would erase the pixels where the text should appear).

There's a bit more going on here. Click the **File** menu and choose **Open**, click **Raspberry Pi Pico**, go into the lib directory, and look for ssd1306.py. Click it, then click **OK** to open it. Scroll down to the line that starts with **class SSD1306_I2C** and look at the first few lines of the **__init__** function:

```
def __init__(self, width, height, i2c, addr=0x3C,
             external_vcc=False):
    self.i2c = i2c
    self.addr = addr
```

The **0x3C** in the first line refers to the address of the I2C device. You can connect many devices to an I2C bus, and each time you want to send or receive data, you need to specify the address of the device you want to communicate with. The **self.addr = addr** line stores the address inside the **SSD1306_I2C** so it can be accessed later.

HEXADECIMAL

Hexadecimal is a base-16 numbering system. That means there are 16 digits: 0–F. So, the number 10 in decimal is A in hexadecimal, and 3C in hexadecimal is 60 in decimal. The advantage of this is that each byte is exactly two digits. This makes it a compact but still understandable way of writing digital information. You'll come across it quite a lot when dealing with I2C and SPI devices.

If you get confused, you can use online hexadecimal-to-decimal converters to switch between the two. For example: **hsmag.cc/hextodec**.

The **write_cmd** function is used inside the library to send commands to the module. Note how it specifies the address as the first argument to **i2c.writeto**:

```
def write_cmd(self, cmd):
    self.temp[0] = 0x80  # Co=1, D/C#=0
    self.temp[1] = cmd
    self.i2c.writeto(self.addr, self.temp)
```

This address is hard-wired into the device (though you may be able to change it on some devices by cutting a trace on the PCB, or soldering a blob — see your device's documentation for details).

You should find the address for any device in its documentation, but you can scan an I2C bus to see what addresses are currently in use. After setting up the I2C bus, you can run the scan method to output the addresses currently in use:

```
import machine
sda=machine.Pin(0)
scl=machine.Pin(1)
i2c=machine.I2C(0,sda=sda, scl=scl, freq=400000)
print(i2c.scan())
```

Of course, there's not much use in a screen that just says Hello World, so let's turn this into something a little more useful — a thermometer. In Chapter 8, *Temperature gauge* you learned how to use the ADC to read temperatures using your Pico's internal temperature sensor. We can now build on this code to make a standalone thermometer that doesn't need a computer to read the output. With your LCD still connected as before, run the following code:

```
import machine
import ssd1306
import time

sda = machine.Pin(0)
scl = machine.Pin(1)
i2c = machine.I2C(0, sda=sda, scl=scl, freq=400000)
display = ssd1306.SSD1306_I2C(128, 32, i2c)

adc = machine.ADC(4)
conversion_factor = 3.3 / (65535)
while True:
    reading = adc.read_u16() * conversion_factor
    temperature = 25 - (reading - 0.706)/0.001721
    display.fill(0)
    display.text(f"Temp: {temperature}", 0, 0, 1)
    display.show()
    time.sleep(2)
```

This should mostly look familiar. The only slight change to the previous temperature code is that before we outputted the result of our calculation — a number — but the OLED needs characters to display, so we use the **str** function which converts the number to a string of characters. We can then build this into a slightly more informative output by combining it with **"Temp: "**. Note that we clear the screen with **fill(0)** so we don't draw new text over the output of previous readings.

As you've seen, I2C is an easy way of linking extra hardware to your Pico. You will need an appropriate library for any device you want to connect, but once you have that, you can easily add all sorts of bits and bobs to your Pico and create impressive builds. You may come across new or exotic hardware that doesn't yet have a library, but if you do, you can usually determine how to work with it using its documentation and by consulting libraries that others have written for similar devices.

Serial Peripheral Interface

We've seen how I2C works, now let's look at SPI. We'll use a slightly different SSD1306 module (a 0.96" 128x64 module that supports SPI), so the commands and everything else are mostly the same, it's just the protocol we send data over that's different.

SPI has four connections: SCK, MOSI, MISO, and CS (sometimes labelled SS). SCK is the clock, MOSI is the line taking data from your Pico to the module, and MISO takes data from a peripheral device to your Pico. CS (Chip Select) and is used to connect many devices to a single SPI bus. You simply take the CS line high to enable an SPI peripheral and pull it low to disable it. In this case of this module, there's an additional reset line (RES) that the library uses to initialise the SSD1306 chip.

Disconnect your Pico from USB, then wire the display as shown in **Figure 10-3**. After it's wired up, reconnect the USB cable so you can program the device. Here's an overview of the connections you need to make:

Raspberry Pi Pico Pin	Breadboard Column	SSD1306 Pin
SCK (GP10 / SPI0 SCK)	14	SCK (or CLK)
MOSI (GP11 / SPI0 TX)	15	SDA (or DIN)
MISO (GP12 / SPI0 RX)	16	DC
CS (GP13 / SPI0 CSn)	17	CS
GP14	19	RES

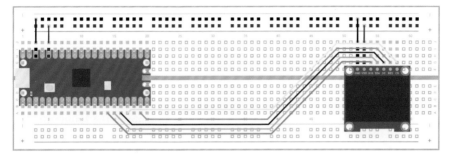

Figure 10-3 Wiring up a 0.96" SSD1306 module for SPI

There are no addresses in SPI, so we can just dive in and write our code:

```
import machine
import ssd1306

mosi = machine.Pin(11)
sck = machine.Pin(10)
res = machine.Pin(14)
dc = machine.Pin(12)
cs = machine.Pin(13)

spi = machine.SPI(1, 100000, mosi=mosi, sck=sck)
display = ssd1306.SSD1306_SPI(128, 64, spi, dc, res, cs)

display.text("Hello World!", 0, 0, 1)
display.show()
```

In this case, we're using SPI1, and one set of available pins for this is GP10, GP11, GP12, and GP13. Most types of serial communication have a speed or baudrate, which is basically how many bits of data it can push through the channel per second. A lot of things affect this, such as the capabilities of the two devices being connected and the wiring between them (how long it is and if there's interference from other devices). If you find you're having problems with mangled data, then you may need to reduce it. For our little screen, we're just sending one byte of data per character, so it doesn't really matter how fast we send it, but for some other SPI devices, fine-tuning the baud rate can be important.

Let's have take a look at how this leaves our thermometer code:

```
import machine
import ssd1306
import time

mosi = machine.Pin(11)
sck = machine.Pin(10)
res = machine.Pin(14)
dc = machine.Pin(12)
cs = machine.Pin(13)

spi = machine.SPI(1, 100000, mosi=mosi, sck=sck)
display = ssd1306.SSD1306_SPI(128, 64, spi, dc, res, cs)

adc = machine.ADC(4)
conversion_factor = 3.3 / (65535)
while True:
    reading = adc.read_u16() * conversion_factor
    temperature = 25 - (reading - 0.706)/0.001721
    display.fill(0)
    display.text(f"Temp: {temperature}", 0, 0, 1)
    display.show()
    time.sleep(2)
```

As you can see, there's really very little difference in the code between I2C and SPI. Once you've got everything set up, the only really change is that with I2C, you must specify the address when you send data, while with SPI you don't (though remember if you had more than one device attached, you'd need to toggle the CS GPIO to select the appropriate device).

So, if they're so similar, which protocol should you choose when building a project? There are a few factors to consider. The first is availability of the things you want to attach. Sometimes a sensor is only available as I2C or SPI, so you have to use that. However, if you've got a choice of hardware, the biggest impact comes when you're using multiple extra devices. With I2C, you can connect as many as 128 devices to a single I2C bus; however, they all need to have a separate address. These addresses are hard-wired in. Sometimes it's possible to change the address with a solderable (or cuttable) connection, but sometimes it's not. If you want to have multiple of the same type of sensors (for example, if you're monitoring the temperature at many points on your project), you may be limited by the number of I2C addresses for your sensor. In this case, SPI may be a better choice.

Alternatively, SPI can have an unlimited number of devices connected; however, each one has to have its own CS line. On the Pico, there are 26 GPIO pins. You need three of them for the SPI bus, so that means there are 23 available for CS lines. And this is assuming you don't need any for anything else. If available GPIOs are at a premium, consider I2C instead.

In reality, for many projects, you can quite happily use either protocol, and you may find that the choice of which to use has more to do with what parts you find in your parts box than a technical difference between the two.

BIT BANGING

Your Pico has two hardware I2C buses and two hardware SPI buses. However, you can use more than these if you want to. Both I2C and SPI can be implemented in software rather than hardware. This means the main processing core handles the communication protocol rather than a specialised bit of the microcontroller. This is known as *bit banging*. While it can be useful, it puts more strain on your processor core than using the specialised hardware, and you may find that you can't reach high baudrates.

The Pico has a trick up its sleeve for this — PIO. We'll look more closely at this later in the book (Appendix C, *Programmable I/O*), but it's an extra bit of hardware in the microcontroller that can be dedicated to input/output protocols such as I2C and SPI. With PIO, you can create extra I2C or SPI buses without taxing the main processor core.

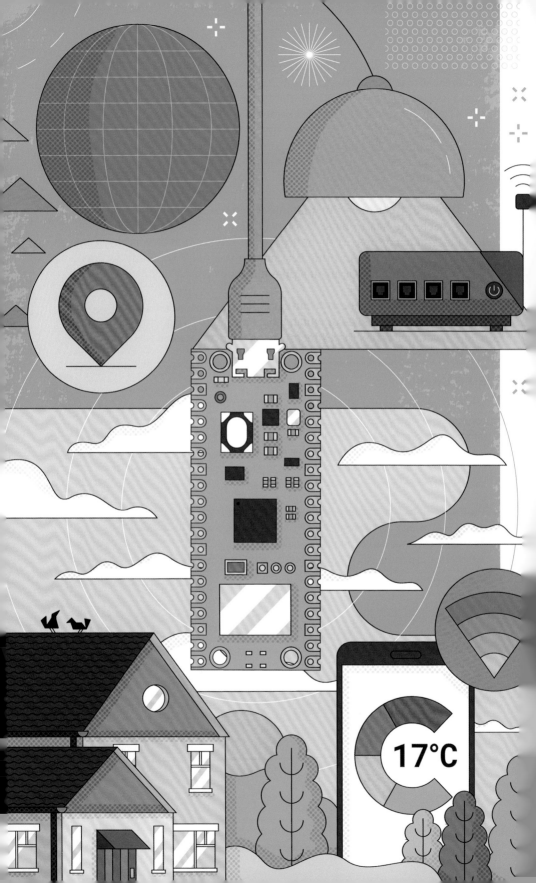

Chapter 11

Wi-Fi connectivity with Pico W

Turn Raspberry Pi Pico W into a network-connected node for the Internet of Things as you learn to unleash its Wi-Fi powers

So far in this book, we've been learning about projects which can be made with either Raspberry Pi Pico or Raspberry Pi Pico W. They're both based on the same RP2040 microcontroller, have the same overall hardware, and even share the same pin connections on the edges of their respective circuit boards. They're not completely identical, though, because Raspberry Pi Pico W has something Pico lacks: a radio.

In this chapter you're going to learn how to make a device for the *Internet of Things (IoT)*, connecting to your home Wi-Fi network. As a result, you'll need Raspberry Pi Pico W; if you're using Raspberry Pi Pico instead, you won't be able to work through this chapter — but read on to learn about what Pico W can do.

The radio

Raspberry Pi Pico W's radio, hidden beneath a metal shield, is designed around two standards: *IEEE 802.11n Wi-Fi*, also known as *Wi-Fi 4,* and *Bluetooth 5.2*. Wi-Fi is a high-speed, relatively long-range wireless networking standard designed for things like smartphones, tablets, and laptops, though it's also popular for embedded devices which need to send or receive large amounts of information. Bluetooth is a short-range, lower-speed radio standard which uses less power, and is used in devices like wireless headphones, smartphones, and sensors. You'll learn about using Bluetooth with Raspberry Pi Pico W in Chapter 12, *Bluetooth connectivity with Pico W*.

Raspberry Pi Pico W's radio is designed to make it suitable for use in projects for the Internet of Things (IoT) — which, like it sounds, is a network made up of devices, rather than people. These devices can be anything from smart thermostats to earthquake monitors, all working to send data off for analysis and receive data in return.

The Internet of Things doesn't have to be complicated, though. In this chapter, you'll learn to harness the power of Pico W's radio to connect to your home network, reach out to other devices on the internet, and even host its own *web server* for you to control Pico W's hardware from a web browser on Raspberry Pi or your smartphone, tablet, or computer.

To complete this project, you'll need:

- Raspberry Pi Pico W — a Pico won't work, as it has no radio

- A Wi-Fi router with an internet connection

- A Raspberry Pi or other computer on the same Wi-Fi network, with a web browser

- The network name (SSID) and password (key) for your Wi-Fi network

Spot the difference

Before programming Pico W to connect to your home network, take a moment to make sure it's running the right firmware. From earlier in the book, you'll remember that there are two distinct versions of the MicroPython firmware: one for Raspberry Pi Pico, and the other for Raspberry Pi Pico W.

If you've installed the Pico firmware on Pico W, it'll seem to work — except you won't be able to use the radio. If you think you might have the wrong firmware installed, simply download the Pico W firmware and flash it using the instructions in Chapter 1, *Get to know your Raspberry Pi Pico*.

MicroPython programs written for Raspberry Pi Pico W are identical to those written for Pico — you can take a program written for Pico and run it on Pico W with no problems. The same isn't directly true the other way around: if a program uses Pico W's radio, it won't work on a Pico.

There's also a small but important difference in the hardware: on Raspberry Pi Pico, the on-board LED is connected to general-purpose input/output (GPIO) pin on the RP2040 microcontroller. Raspberry Pi Pico W, though, had to use this GPIO pin to communicate with the radio — but it still has an on-board LED.

Rather than being controlled by the RP2040, though, this LED is controlled by the radio controller. You'll read more about controlling Pico W's on-board LED later in this chapter. For now, just be aware of the difference — and if you've got programs written for Pico which address the LED by pin number and aren't working on Pico W, now you know why!

Making a connection

To connect Raspberry Pi Pico W to your Wi-Fi network, you'll need two pieces of information: the network name, also known as the *Service Set Identifier (SSID)*, and the password, also known as the *Pre-Shared Key (PSK)*. This information is often written on a sticker or card attached to your router or access point, unless it has been changed from the factory defaults, and controls who can access the network.

The radio on Pico W is what is known as a *single-band radio*, meaning it only uses one block of frequencies in the radio spectrum: 2.4GHz. As a result, it can't connect to 5GHz or 6GHz Wi-Fi networks. If your router has separate network names for each band, make sure to use the name associated with the 2.4GHz band; if it uses one network name for all bands, Pico W will still connect — but it may take longer than if you were using a single-band network.

MicroPython programs for Pico W start like any other: open Thonny and start a new program, but this time you'll be importing a library you haven't used before: the **network** library. Type the following:

```
import time
import network
import rp2
rp2.country("GB")

wlan = network.WLAN(network.STA_IF)
wlan.active(True)
wlan.connect("NetworkName", "Password")
```

Here you're importing the **network** library to handle Pico W's radio, the **time** library to handle delays, and the **rp2** library to handle the RP2040 microcontroller. The line after the import section is a function of the **rp2** library that sets the *country code* for the radio. If you've ever set up a Raspberry Pi, you'll be familiar with the country code setting from the welcome wizard: different countries have different regulations when it comes to radio frequencies, so you need to tell your device where it is in the world to use all permitted radio frequencies and Wi-Fi channels.

In this example you're setting the country code to **GB**, which is correct for the United Kingdom. If you're elsewhere in the world, you'll need to look up the right country code for your nation using what is known as the *ISO 3166 Alpha-2* format. For Ireland, it's **IE**; for the United States, it's **US**; for Canada, it's **CA**. A search engine will give you a full list.

WARNING

Make sure you use the correct country code, or leave it unset. Using an incorrect country code can mean Pico W won't be able to see your wireless network, or — worse — cause it to transmit on unauthorised frequencies, causing interference and potentially breaking laws surrounding radio use in your country.

The last three lines in your program need some explanation: here you're creating an object called **wlan** — short for Wireless Local Area Network — which uses the **network** library to set Pico W's radio into *station mode*. A station is simply any device which connects to the network, like a smartphone or Raspberry Pi. The radio can also be brought up in *access point mode* using **network.WLAN(network.STA_IF)** to allow other devices to connect directly to Pico W without a router in the middle; you won't be using this mode in this book, though.

The second of the three lines tells the **network** library to turn Pico W's radio on, while the last instructs it to connect to a network with the name **Network-Name** and the password **Password**. Obviously, you'll need to change these to match your own network name and password — or your Pico W won't connect!

To check the status of the connection, write a simple loop:

```
while not wlan.isconnected() and wlan.status() >= 0:
    print("Waiting for Wi-Fi connection...")
    time.sleep(1)
print(wlan.ifconfig())
print(wlan.isconnected())
```

Here, you're using the **network** library's **isconnected()** function, along with the **status()** function, to keep the loop running only when Pico W is *not* connected to your network. When it does connect, and the radio doesn't report any error status, the loop will exit. Finally, MicroPython will print out information about the connection, including the *IP address* assigned to Pico W — the network equivalent of its telephone number or street address.

Your finished program should look like this:

```
import time
import network
import rp2
rp2.country("GB")

wlan = network.WLAN(network.STA_IF)
wlan.active(True)
wlan.connect("NetworkName", "Password")

while not wlan.isconnected() and wlan.status() >= 0:
    print("Waiting for Wi-Fi connection...")
    time.sleep(1)
print(wlan.ifconfig())
print(wlan.isconnected())
```

Save the program to your Pico W as **Network_Connect.py** and click Run. You'll see the message 'Waiting for Wi-Fi connection...' in the Shell area once per second until Pico W connects, then a list of the radio's configuration. If the last line of the output is **True**, congratulations: you're online!

NO CONNECTION?

If Pico W hasn't connected to your network after about a minute, there's something wrong. Check that you wrote the network name and password correctly, and that you're connecting to a 2.4GHz network — not a 5GHz or 6GHz network. Also double-check that you've set the correct country code — remember that the United Kingdom's code is 'GB', not 'UK'!

That simple program is enough to get you connected, but it's not exactly robust. Using the **network** library's ability to report the radio's status, including errors, it's possible to add *error handling* to your program — allowing it to let you know when there's a problem, and even what type of problem it might be. Go back to Thonny and rewrite your program as follows:

```
import time
import network
import rp2
rp2.country("GB")

ssid = "NetworkName"
psk = "Password"

wlan = network.WLAN(network.STA_IF)
wlan.active(True)
wlan.connect(ssid, psk)
```

```
max_wait = 30
while max_wait > 0:
    if wlan.status() < 0 or wlan.status() >= 3:
        break
    max_wait -= 1
    print("Waiting for Wi-Fi connection...")
    time.sleep(1)

if wlan.status() != 3:
    raise RuntimeError("Network connection failed")
else:
    print("Connected to Wi-Fi network.")
    print(wlan.ifconfig())
```

Save your program as **Connect_Robust.py** and click **Run**. This time, your Pico W will begin the connection attempt and then count backwards from 30 — the **max_wait** variable — as it checks to see if the connection has succeeded yet. When it reaches zero, or the connection succeeds, it will exit the loop and either print a success message and the network configuration information from your first program, or it will *raise an error* — letting you know that the connection didn't work.

The program relies on the **network** library's ability to report the status of Pico W's radio, but it comes through in a format designed for a machine to understand: a simple number. A status of '3' means that the radio has successfully connected to the network, which is why your program raises the error if the loop has exited but the status is not 3.

You may notice one other change to the program: the network name and password are now stored in variables, **ssid** and **psk**. This is a good habit to build, as it makes it easier to see where you need to enter new network details when sharing the program — and makes it less likely you'll accidentally share the program with your own network password still in it!

STATUS CODES

There are other status numbers, which you can watch for in your program using the **network** library's **status()** function if you want more detailed error reports: 0 means the connection is down; 1 means the radio is currently joining a network; 2 means the radio connected but was not given an IP address by the router; -1 means the radio link has failed; -2 means the radio is connected but there's no underlying network; and -3 means that the network password was rejected.

Connecting to the internet

Now you know how to handle connecting to the network, it's time to get Pico W to actually use the network — starting with talking to web servers on the internet. Although Pico W isn't powerful enough to run a graphical web browser like Raspberry Pi or a desktop computer, it can still talk to the same servers, using what is known as the *Hypertext Transport Protocol (HTTP)*.

There's another MicroPython library which exists specifically to request data from web servers: **requests**. Go back to the import section of your program and add the following line:

```
import requests
```

Now go to the very bottom of your program and add the following lines:

```
response = requests.get("https://text.npr.org")
print(response.content)
response.close()
```

Here we've created a **requests** object, called **response**, which contains the full address — including the *protocol*, the **https://** bit — of the website we want Pico W to visit. In this example, you're opening a secure connection to the website — that's what the 's' in **https://** means — but you can also connect to plain **http://** servers too.

The next line is simple enough: it tells Pico W to print the content it received from the web server to the Shell area. Don't expect it to look much like it would in a normal web browser, though: you won't see any pictures, pretty fonts, or animations. Instead, you'll see the raw text of the response — the Hypertext, in fact.

Your finished program should look like this:

```
import time
import network
import rp2
import requests
rp2.country("GB")

ssid = "NetworkName"
psk = "Password"

wlan = network.WLAN(network.STA_IF)
wlan.active(True)
```

```
wlan.connect(ssid, psk)

max_wait = 30
while max_wait > 0:
    if wlan.status() < 0 or wlan.status() >= 3:
        break
    max_wait -= 1
    print("Waiting for Wi-Fi connection...")
    time.sleep(1)

if wlan.status() != 3:
    raise RuntimeError("Network connection failed")
else:
    print("Connected to Wi-Fi network.")
    print(wlan.ifconfig())

response = requests.get("https://text.npr.org/")
print(response.content)
response.close()
```

Save the program to your Pico as **Requests.py** and run it now. Your Pico W will connect to the network, as before, then make a request to a lightweight version of the NPR website. When it receives the page back, it prints it to the Shell area; click the grey "squeezed" text to see all of it, and put a cross in the **Wrap text** box to make it easier to read, as shown in **Figure 11-1**.

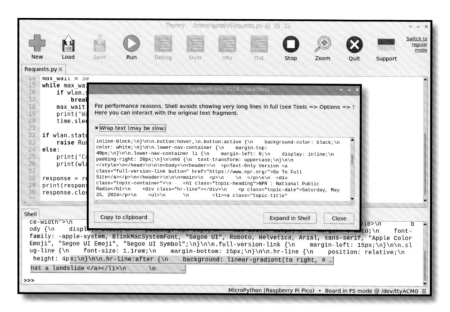

Figure 11-1 Unsqueezing Shell area output in Thonny

For something easier to read, you can tell MicroPython to split the response into separate lines according to the line breaks it receives from the server. To do that, delete your existing **print** line and replace it with:

```
[print(x) for x in response.content.splitlines()]
```

If you save and run your program again, you'll see that the response from the server is spread out more — making it better for reading, but still not as it would appear if processed and rendered as rich text in a standard web browser. The separate lines are generated by reading each split line from the **response** object and printing them one at a time, until they're all finished.

All that happens before the final line, and your program would seem to work without it — for a while, at least. The function **response.close()** tells **requests** that you're finished requesting data from the server, and to close the *socket* (the connection between Pico W and the web server) it opened.

You need to make sure that every time you use **requests** to get content from a remote server you include a **response.close()** function. If you don't, all the connections will stay open — and Pico W will very quickly run out of memory, causing your program to crash. When the socket is closed, a process called *garbage collection* takes place to free up the memory that was previously being used — allowing your program to keep running as long as you need.

OUT OF MEMORY?

Even if you're making sure to close your requests each time, you may find yourself running out of memory. Modern web pages are big, even if you're not downloading the pictures, and Pico W has just 264 kB of memory. If you request a page that's too big, your program will crash; try to find a site which offers the information you need in a smaller page.

The NPR website looks great in a browser, but not so good on the Pico W. You can pull data from any website, though, including ones which are designed for providing useful information on devices that can't handle graphical pages. Try replacing your request with one of the following examples and running your program again, to see what you get:

```
response = requests.get("http://wttr.in/cambridge?format=3")
response = requests.get("http://ipecho.net/plain")
response = requests.get("https://earthquake.usgs.gov/fdsnws/event/"
                        "1/query?format=text&limit=10")
response = requests.get("http://artscene.textfiles.com/"
                        "asciiart/unicorn")
```

In addition to requesting content from a web server — using what is known as an *HTTP GET request* — you can also send data to a web server, using an *HTTP POST request*. Like the name suggests, this *posts* data to the server rather than *getting* data from it — allowing you to send things like sensor readings across the network.

For this you'd need what's known as an *endpoint* set up to accept the data and do something with it, like store it in a database; that's outside the scope of what you're doing in this chapter, but you can find more information on making POST requests in the free book *Connecting to the Internet with Raspberry Pi Pico W* on the Raspberry Pi website at **rptl.io/picow-connect**.

Hosting a web page

Pico W isn't limited to just talking to other peoples' web servers: it can become a web server itself, too. While admittedly limited by its memory and storage capacity, a Pico W-hosted server can prove extremely handy — especially when you start using it to interact with sensors and other hardware.

Start a new program in Thonny, and begin by configuring the network — using the same tricks as before to set the connection up in a way that it won't stall forever if it's having trouble connecting. To make it easier, you can literally just copy and paste all the lines from your previous program then delete the last three as well as the line **import requests** near the top — leaving you with a program which connects Pico W to the Wi-Fi network but does nothing else.

Save your program on Pico W as **connect.py**. You can run it to check that everything's working, but it won't do anything beyond connect Pico W to your Wi-Fi network and print the connection status.

Now start another new program. Don't worry, you'll still be using what you've just written — only this time you'll be importing it, like a library, so that you don't have to write it out in full every time you're using Wi-Fi on Pico W. You'll be importing a couple of additional libraries, too: **socket** and **machine**. Type the following:

```
from connect import wlan
import socket
import machine

address = socket.getaddrinfo("0.0.0.0", 80)[0][-1]
s = socket.socket()
s.setsockopt(socket.SOL_SOCKET, socket.SO_REUSEADDR, 1)
s.bind(address)
```

```
s.listen(1)
print("Listening for connections on", wlan.ifconfig()[0])
```

This tells Pico W to use **connect.py** to connect to the Wi-Fi network, then open a network socket on port 80 — the port used for HTTP-protocol connections from web browsers and other devices. The socket is linked to the Pico W's network address, and then told to listen on it for connections. Save your program on Pico W as **server.py**, but don't run it: if you ran your program now you wouldn't see anything, as there's no content for Pico W to serve.

To fix that, add the following:

```
while True:
    try:
        client, address = s.accept()
        print("Connection accepted from", address)
        client_file = client.makefile("rwb", 0)
        while True:
            line = client_file.readline()
            if not line or line == b"\r\n":
                break
        client.send("HTTP/1.0 200 OK\r\n")
        client.send("Content-type: text/plain\r\n\r\n")
        client.send("Hello from Raspberry Pi Pico W!\r\n")
        client.close()
        print("Response sent, connection closed.")
    except OSError as e:
        client.close()
        print("Error, connection closed.")
```

This code sets up a loop, in which the **socket** library accepts incoming connections from other devices on the network — using an object called **client**, to indicate that you're dealing with clients for the web server you're creating. When a connection is requested, a line is printed to the Shell area and then a temporary file created to hold the client request — read line-by-line in the nested loop.

The **client** object is then used to **send** a response — starting with a bit of boilerplate which tells the client what version of the Hypertext Transport Protocol you're using (version 1.0, here), that the server has accepted the connection (**200 OK**), and that you're sending content of the type **text/plain** — we'll look at another content type later. The **\r\n** on the end of the response sends a carriage return followed by a newline, special characters which let the receiving computer know the line has ended.

The second `client.send` line sends your actual message — a simple network 'hello, world,' in effect. The first two lines of the response won't be seen in the connecting device's web browser, unless you're using special software to inspect it; the third line, though, will be printed in your browser. Prove it: save and run your program now. You'll need to press the **Stop** button first if you'd already tried running it: even though there was nothing to serve, Pico W is still busy in the loop and needs to be told to stop running the program before you can save your changes.

When the program has connected to the network, you'll see a line printed to the Shell area which starts with 'Listening for connections on' followed by an IP address — four numbers separated by dots. This is how you're going to connect to your Pico W over the network, like calling a friend using their telephone number.

Open a web browser and type the IP address from the Shell area into the address bar. A very simple page will load, containing the message 'Hello from Raspberry Pi Pico W!' Congratulations: you've built a web server, running on your Pico W! You don't have to connect to it from the same computer on which you're programming the Pico W, either: so long as they're on the same network, any device — from a desktop or laptop to a smartphone, tablet, or even a games console — will be able to load the page using that IP address.

> **❗ WARNING**
>
> The IP address you're using is what's known as a *private address*. It can only be accessed by people on the same network as you. The websites you usually visit have *public addresses*, which you can access over the internet. If you send the IP of your Pico W to a friend down the street, they won't be able to load your page. It's possible to forward connections from the public address assigned to your router to the private address assigned to Pico W, but it's not recommended unless you're sure about what you're doing: forwarding the connections means anyone can access your device.

Just having a static message isn't using the full functionality of Pico W, though. It's time to use the `machine` library you imported earlier in your code to interface with Pico W's hardware — specifically, to read the temperature sensor. The code you need for this is the same as in Chapter 8, *Temperature gauge*.

Start by adding the temperature sensor to your program, along with the conversion factor for turning its readings into a voltage value for later processing — put the following lines at the top of your program, under `import machine`:

```
sensor_temp = machine.ADC(4)
conversion_factor = 3.3 / (65535)
```

Next, go down to your main program loop and add the following under the line **try:**

```
reading = sensor_temp.read_u16() * conversion_factor
temperature = 27 - (reading - 0.706) / 0.001721
```

Finally, the following new lines between your last **client.send** and **client.close**:

```
response = f"The temperature is {str(temperature)} C.\r\n"
client.send(response)
```

The first of these two lines constructs an object called response, which joins together a couple of descriptive strings — and the carriage return and newline characters required at the end of each line — with the converted reading from Pico W's temperature sensor, converted from its floating-point number data type to a string. The second then sends the result to the client device.

Your finished program should now look like this:

```
from connect import wlan
import socket
import machine

sensor_temp = machine.ADC(4)
conversion_factor = 3.3 / (65535)

address = socket.getaddrinfo("0.0.0.0", 80)[0][-1]
s = socket.socket()
s.setsockopt(socket.SOL_SOCKET, socket.SO_REUSEADDR, 1)
s.bind(address)
s.listen(1)
print("Listening for connections on", wlan.ifconfig()[0])

while True:
    try:
        reading = sensor_temp.read_u16() * conversion_factor
        temperature = 27 - (reading - 0.706) / 0.001721
        client, address = s.accept()
        print("Connection accepted from", address)
        client_file = client.makefile("rwb", 0)
        while True:
            line = client_file.readline()
            if not line or line == b"\r\n":
                break
```

```
    client.send("HTTP/1.0 200 OK\r\n")
    client.send("Content-type: text/plain\r\n\r\n")
    client.send("Hello from Raspberry Pi Pico W!\r\n")
    response = f"The temperature is {str(temperature)} C.\r\n"
    client.send(response)
    client.close()
    print("Response sent, connection closed.")
except OSError as e:
    client.close()
    print("Error, connection closed.")
```

Save your program as **Temperature_Server.py** on your Pico W and run it —
remembering to press the Stop button first — and refresh the page in your browser. Now beneath the welcoming message you saw before is a new line — with a
live reading from the temperature sensor. Place your finger on the RP2040 microcontroller chip for a few seconds to increase its temperature, then refresh
the page again — and you'll see the number change.

Congratulations: you've created a dynamic page which reads from Pico W's
temperature sensor!

Controlling an LED

Reading sensors over a Wi-Fi connection is neat, but Pico W's capabilities
extend beyond that: it's possible to actively control hardware as well. For
this project, you'll be using the Pico W's on-board LED; if you'd prefer to use
something connected to a GPIO pin instead, like a bigger external LED, a motor, or a buzzer, follow the instructions earlier in this book to modify what pin
the program controls.

Start a new program, importing the libraries you'll need plus your **connect.py**
code — only this time make sure you import the **machine** library and set the
on-board LED up as an output:

```
from connect import wlan
import socket
import machine

led_onboard = machine.Pin("LED", machine.Pin.OUT)
led_onboard.value(0)
led_state = "LED is off"
```

Note how you're setting the pin up based on a label, **"LED"**, rather than a
pin number. You might remember from earlier in the chapter that, unlike

Raspberry Pi Pico, Pico W doesn't control the on-board LED from one of the RP2040 microcontroller's pins but from the Wi-Fi controller instead. The **"LED"** label is special, and lets MicroPython know which is the right pin whether you're using Pico or Pico W without you having to make any changes to your program between the two. The line beneath it, meanwhile, tracks whether the LED is on or off — your program explicitly turns the LED off when it starts, so you can set the state-tracking variable to off.

This time, Pico W isn't going to be sending your web browser a plain text response; instead, it's going to send Hypertext Markup Language (HTML), which is what lets web pages control their formatting and include hyperlinks and other interactive features. To make that easier to manage, create a variable to hold the page contents — but note that you'll have to indent the code yourself, as Thonny won't create the indentations automatically:

```
html = """
<!DOCTYPE html>
<html>
    <head> <title>Raspberry Pi Pico W</title> </head>
    <body> <h1>Raspberry Pi Pico W</h1>
        <p>%s</p>
    </body>
</html>
"""
```

Here, you're building the skeleton of an HTML-format document — including a **<head>** section with a page title and a **<body>** section with a heading and a placeholder. You'll notice that every *HTML tag* that's opened has a matching tag with a slash at the front: this closes the tag, like telling a word processor to stop making what you're typing bold or italicised.

The **%s** section is special: this is a placeholder for *dynamic content* — in other words, page content which will change depending on what you're doing. You'll see what this is for later in the program.

Next, open a socket the same way as in your previous program:

```
address = socket.getaddrinfo("0.0.0.0", 80)[0][-1]
s = socket.socket()
s.setsockopt(socket.SOL_SOCKET, socket.SO_REUSEADDR, 1)
s.bind(address)
s.listen(1)
print("Listening for connections on", wlan.ifconfig()[0])
```

Then tell Pico W to listen not only for connections from clients, but for specific *requests*:

```
while True:
    try:
        client, address = s.accept()
        print("Connection accepted from", address)
        request = client.recv(1024).decode("UTF-8")
        print(request)
```

The request is how you're going to control the LED, using what is known as a *representational state transfer application programming interface*, or *RESTful API*. That sounds like something complicated, but it's actually much simpler than it sounds: an application programming interface simply provides a way for a program to talk to something, and in this case, it follows a set standard for sending and receiving information about the state of objects. Technically speaking, what you're writing here isn't a true RESTful API — but it's a good introduction to the core concepts.

You need a way to handle incoming requests from the client, so add that next:

```
led_on = request.startswith("GET /led/on")
led_off = request.startswith("GET /led/off")
print("led_on = " + str(led_on))
print("led_off = " + str(led_off))
```

Here you're searching through the request that came from the client for two strings: **/led/on** and **/led/off**. These will act as toggles for switching the LED on and off, exactly as it looks — but you're going to need more code for that to actually happen:

```
if led_on:
    print("Client requested to turn the LED on.")
    led_onboard.value(1)
    led_state = "LED is on"

if led_off:
    print("Client requested to turn the LED off.")
    led_onboard.value(0)
    led_state = "LED is off"
```

Here you're checking to see whether your search through the request found the strings **/led/on** or **/led/off**, and changing the value of the LED accordingly — turning it on or off, exactly as the client requested. If neither string was found in the request, then the LED is left alone — off if it was already off, or on if it was already on. Your program is also updating the **led_state** variable, changing it to on or off as required.

Finally, build the response using the HTML skeleton you built earlier in the program and the **led_state** variable, and send it to the client:

```
response = html % led_state

client.send("HTTP/1.0 200 OK\r\n")
client.send("Content-type: text/html\r\n\r\n")
client.send(response)
client.close()

except OSError as e:
    client.close()
    print("Error, connection closed.")
```

Your finished program should look like this:

```
from connect import wlan
import socket
import machine

led_onboard = machine.Pin("LED", machine.Pin.OUT)
led_onboard.value(0)
led_state = "LED is off"

html = """
<!DOCTYPE html>
<html>
    <head> <title>Raspberry Pi Pico W</title> </head>
    <body> <h1>Raspberry Pi Pico W</h1>
        <p>%s</p>
    </body>
</html>
"""

address = socket.getaddrinfo("0.0.0.0", 80)[0][-1]
s = socket.socket()
s.setsockopt(socket.SOL_SOCKET, socket.SO_REUSEADDR, 1)
s.bind(address)
s.listen(1)
print("Listening for connections on", wlan.ifconfig()[0])

while True:
    try:
        client, address = s.accept()
        print("Connection accepted from", address)
        request = client.recv(1024).decode("UTF-8")
```

```
    print(request)

    led_on = request.startswith("GET /led/on")
    led_off = request.startswith("GET /led/off")
    print("led_on = " + str(led_on))
    print("led_off = " + str(led_off))

    if led_on:
        print("Client requested to turn the LED on.")
        led_onboard.value(1)
        led_state = "LED is on"

    if led_off:
        print("Client requested to turn the LED off.")
        led_onboard.value(0)
        led_state = "LED is off"

    response = html % led_state

    client.send("HTTP/1.0 200 OK\r\n")
    client.send("Content-type: text/html\r\n\r\n")
    client.send(response)
    client.close()

except OSError as e:
    client.close()
    print("Error, connection closed.")
```

Save and run your program now. When Pico W has connected to the network, open a web browser and type its IP address into the address bar. You'll see a simple web page load — looking considerably prettier than the plain-text response of your earlier program — which tells you what you probably already know: the on-board LED is switched off.

To control the led, you need to construct your request — which is as easy as appending **/led/on** to the address. For example, if Pico W has the IP address **192.168.50.10**, you'd type **192.168.50.10/led/on** into the browser's address bar.

When you load that page, you'll see Pico W's on-board LED light up. If you look at the response you receive in the browser, you'll also notice that the dynamic section of the page — the **%s** placeholder — has been updated, and now says 'LED is on' instead of 'LED is off'. To turn the LED off again, simply use **/led/off** instead. To check the current status of the LED — assuming you can't just look at Pico W and see for yourself, of course — just type the IP address without anything after it.

Having to append your request to the end of the address manually is a pain, however — but one you can alleviate by adding an interactive button to the page. Go back to the top of your program and find the HTML skeleton you built, then below the `<p>%s</p>` line add the following simple form:

```
<form action="%s">
    <input type="submit" value="%s" />
</form>
```

To have the button provide a useful toggle, you need to know whether the LED is currently on or off. Find your two **if** statements, and add the following beneath the second:

```
if led_state == "LED is on":
    button_link = "/led/off"
    button_text = "Turn LED off"

else:
    button_link = "/led/on"
    button_text = "Turn LED on"
```

This will create a button to turn the LED on if it's off, or off if it's on. Finally, you'll need to add these two new variables to the response you're building so that the form appears in the page correctly. Find the **response = line** and change it to read:

```
response = html % (led_state,button_link,button_text)
```

Save your program — remembering to click the **Stop** button if it was already running — then click **Run** and reload the page in your browser. You'll see a new button has appeared: click it, and Pico W's on-board LED should turn on. Click it again, and it'll turn off. Keep clicking it, and you can create a very small disco light show.

Congratulations: you can now control Pico W over the network! From here, you can build on these concepts to create more complicated projects: how about a web page that lets you know if you've left a door open, or that lets you turn on a light? Try connecting motors to Pico W, using a suitable motor driver, and making a simple robot you can control from your browser.

To learn about Pico W's more advanced Wi-Fi capabilities, including its ability to process data in a format known as *JSON* and to submit data to a remote endpoint server, read more in *Connecting to the Internet with Raspberry Pi Pico W* (**rptl.io/picow-connect**).

Chapter 12

Bluetooth connectivity with Pico W

Link Raspberry Pi Pico W to your smartphone, tablet, or another Pico W with Bluetooth Low Energy

Designed for short-range wireless communication, Bluetooth is a technology you likely use every day without realising it: games console controllers typically connect via Bluetooth, as do wireless headphones and mobile phone headsets, and many wireless mice and keyboard use Bluetooth too.

The radio in Raspberry Pi Pico W supports Bluetooth as well as Wi-Fi, in two variants: *Bluetooth Classic* and *Bluetooth Low Energy* (*BLE*). BLE offers a more energy-efficient take on wireless communication, and is frequently used in the Internet of Things (IoT) to gather data from sensors — which is exactly what you'll be doing in this chapter.

To complete this project, you'll need:

▸   One or two Raspberry Pi Pico W boards

▸   A smartphone or tablet with Bluetooth Low Energy (BLE) support

▸   The Punch Through Design LightBlue app, available free for Apple iOS and Google Android from their respective app stores.

About Bluetooth

First released in 1998 by the Bluetooth Special Interest Group, Bluetooth has a much shorter range than Wi-Fi and transfers data at a much slower rate —

but using much less energy, making it ideal for battery-powered and embedded devices.

The original Bluetooth standard, now known as Bluetooth Classic, was joined in 2009 by Bluetooth Low Energy (BLE). As the name implies, Bluetooth Low Energy requires even less energy than Bluetooth Classic — making it the go-to standard for connecting embedded systems like sensors.

Although the radio in Pico W supports both Bluetooth Classic and Bluetooth Low Energy modes, at the time of writing, only Bluetooth Low Energy was supported in MicroPython — and it's the version you'll use in this chapter.

A BLE temperature sensor

Like in Chapter 11, *Wi-Fi connectivity with Pico W*, using a Bluetooth Low Energy connection on Pico W requires quite a lot of 'boilerplate' code — static MicroPython code for activating the radio, defining functions for transmitting *advertising beacons*, decoding messages, and the like, which doesn't change from program to program.

ADVERTISING, NOT ADVERTS

In this context, 'advertising' doesn't mean Pico W is going to try to sell you something. Rather, it simply means that it will broadcast a message to any nearby devices telling them what services it offers — thus 'advertising' its capabilities.

Thankfully, there's an easier way: the **aioble** library. Designed for use along with MicroPython's **bluetooth** library, **aioble** handles a lot of the boilerplate work for you — making it programs which use it shorter and easier to understand than their equivalents written using the **bluetooth** library alone.

The programs in this chapter build on the examples provided with the **aioble** library. To see the original examples, library documentation, and additional examples, head to the aioble GitHub repository at **rptl.io/aioble**.

Open Thonny and create a new file, then type in the following:

```
import struct
import asyncio
import aioble
import bluetooth
```

This imports the four libraries you'll be using. **aoible** and **bluetooth** are for handling the Bluetooth Low Energy work; **struct** is for handling structured data, which you'll be using to create a *payload* for your BLE advertising beacon; and **asyncio** is a library for handling *asynchronous tasks*.

An asynchronous task is a little like a thread, which you used in Chapter 5, *Traffic light controller* to monitor a push-button switch without tying up the main program thread. Where Pico W has two cores to run up to two threads simultaneously, though, it can run as many asynchronous tasks as you like — though, as it's technically switching back and forth very quickly between them, if you run too many at once performance will suffer.

Because you'll be using the Pico W's built-in temperature sensor in this project, you'll need to set things up the same as in Chapter 8, *Temperature gauge*. Add the following two lines to your program:

```
sensor_temp = machine.ADC(4)
conversion_factor = 3.3 / (65535)
```

These, you may remember, tell Pico W to use its fourth analogue-to-digital converter (ADC), connected to the RP2040's temperature sensor, and set up a mathematical conversion to map the read value so it can be converted into a temperature later in the program.

Next, you'll need to set up some variables for your BLE advertising beacon. Add the following lines:

```
ble_name = "picow_ble"
ble_svc_uuid = bluetooth.UUID(0x181A)
ble_characteristic_uuid = bluetooth.UUID(0x2A6E)
ble_appearance = 0x0300
ble_advertising_interval = 2000
```

The variable **ble_name** is reasonable straightforward: it's a friendly name given to identify your beacon. It doesn't have to be **picow_ble** — you could call it **kitchen_temperature**, **temp_sensor**, **sara**, or any other name of your choice.

The remaining lines, though, should be left as they are — for now, at least. The first sets a *universally unique identifier* (*UUID*) for the beacon's *service type*. This isn't chosen at random, unlike the name, but comes from a pre-set list of possible values set by the Bluetooth Special Interest Group (Bluetooth SIG), which is in charge of the Bluetooth and Bluetooth Low Energy (BLE) standards. Here, you're setting the UUID to the hexadecimal number 0x181A, which is 6,170 in decimal. That corresponds to a service type of 'environmental sensing," meaning a device which measures some aspect of its surrounding environment.

The line below it sets another UUID, this time for the beacon's *characteristic*: 0x2A6E, 10,862 in hexadecimal, which is the UUID for a temperature sensor. Put together, the two UUIDs mean that the beacon you're creating will be an environmental sensor which measures temperatures.

ble_appearance isn't a UUID, but a hexadecimal value which chooses from a list of icons to represent the beacon. In this case, it's icon 0x0300, or 768 in decimal, which is a generic picture of a thermometer. These icons are only displayed on certain devices, and you won't see them on Pico W itself, but you should set one anyway for completeness. Finally, the *advertising interval* is the time, in milliseconds, between your beacon's advertising broadcasts.

KNOW YOUR UUIDS

These values are fine for the beacon you're making now, but you'll need to change them if you want to build anything other than a temperature sensor beacon. A full list of UUIDs is available in the Bluetooth SIG's Assigned Numbers Document — totalling nearly 400 pages and covering every possible service, characteristic, and appearance option — at **rptl.io/btnums**.

You'll now need to do a little more setup for the BLE beacon part of the program. Add the following lines:

```
ble_service = aioble.Service(ble_svc_uuid)
ble_characteristic = aioble.Characteristic(
    ble_service,
    ble_characteristic_uuid,
    read=True,
    notify=True)
aioble.register_services(ble_service)
```

For the line starting **ble_characteristic**, you can press the **ENTER** key after the opening bracket and after every comma to make it easier to read, as shown — Thonny will automatically indent your code for you. Alternatively, you can reduce the number of lines in the program by writing it all on one line — finishing at **notify=True)**.

Here, you're setting up **aioble** with the details you inserted earlier in your program. If you want to change them later — to choose a different name, or to change the service or characteristic type, you don't need to touch this block of code at all.

The two most important sections are **read=True** and **notify=True**, which configure your BLE beacon so that its values can be read — because it wouldn't be much use if they couldn't — and so that devices can choose to *subscribe* to the beacon to receive notifications when new readings are available.

Next, you need to create the tasks which will run side-by-side to make the beacon work. The first one you'll need is the beacon task itself. Add the following lines to your program:

```
async def ble_task():
    while True:
        async with await aioble.advertise(
            ble_advertising_interval,
            name=ble_name,
            services=[ble_svc_uuid],
            appearance=ble_appearance) as connection:
            print("Connection from", connection.device)
            await connection.disconnected()
```

Here you're creating an *asynchronous coroutine* which uses **aioble** to activate an advertising beacon, transmit at the interval chosen earlier, and to print a message to the shell area when a device connects. This block of code won't run right now, but only when it's called as a task later in the program.

Before you can create your next coroutine, you need to create a *helper function* which will take the temperature readings from the sensor and format them in the way that **aioble** needs in order to transmit them as a payload in the advertising beacon. Add the following two lines:

```
def encode_temp(temperature):
    return struct.pack("<h", int(temperature * 100))
```

This function takes the temperature reading from the sensor and turns it into *structured data* in the format expected of a BLE beacon. It also converts it from a floating-point number — one with a decimal point — to an integer, multiplying it by 100 first. That multiplication will be important later, when you come to read the payload, so remember it!

With the helper function set up, you can create your second asynchronous coroutine — responsible for actually reading the temperature sensor. If you've worked through Chapter 9, *Data logger*, the next few lines will look very familiar:

```
async def sensor_task():
    while True:
```

```
reading = sensor_temp.read_u16() * conversion_factor
temperature = 27 - (reading - 0.706) / 0.001721
print("Temperature:", temperature)
ble_characteristic.write(encode_temp(temperature))
await asyncio.sleep_ms(2000)
```

Here, you're reading the value from the temperature sensor via the fourth analogue-to-digital converter (ADC) and multiplying it by the conversion factor, then using the 'magic numbers' for the RP2040 to convert that reading to a temperature in degrees Celsius. This is then printed to the shell, encoded using the helper function you just wrote, and written as a BLE beacon payload.

Finally, the coroutine goes to sleep for 2,000 milliseconds — two seconds. It doesn't use the normal **sleep()** instruction, though: here you're using a special version which comes with the **asyncio** library, as it's *non-blocking*. The normal **sleep()** instruction is *blocking*, meaning that your program can't do anything else while it's sleeping; **asyncio.sleep_ms()** isn't, and the other task in your program can continue to run even while your sensor task is sleeping until its next scheduled reading.

Like your earlier coroutine, this block of code won't run until the task is started. To do that, add the following lines to your program:

```
async def main():
    task1 = asyncio.create_task(ble_task())
    task2 = asyncio.create_task(sensor_task())
    await asyncio.gather(task1, task2)
```

This creates two tasks, named **task1** and **task2**, corresponding to the BLE beacon coroutine and the sensor reading coroutine respectively. The two tasks are then set running using **asyncio.gather()**, while **await** means that the main coroutine will continue to run until both tasks have finished. Because both tasks are based on coroutines which loop infinitely, that means the main coroutine will never finish — keeping your beacon running so long as Pico W has power.

Even now, though, your code won't actually run — but your almost there. Add the final two lines as shown:

```
print("Launching Raspberry Pi Pico W BLE temperature sensor...")
asyncio.run(main())
```

This prints a message to the shell and tells the asyncio library to run its main coroutine, converting the two other coroutines into tasks and setting them running. That's it: your program is finished!

Save this program to your Pico W as **BLE_Temperature.py** and click **Run**. You'll see your launch message printed to the shell area, followed by a temperature reading every two seconds. To read the transmitted beacon, though, you'll need a device capable of receiving Bluetooth Low Energy beacons.

The good news is, you probably already do: all modern smartphones and most tablets include Bluetooth radios with Bluetooth Low Energy support as standard. The software you'll need is free, too: open the Apple App Store or Google Play, on iOS or Android respectively, and search for the 'Punch Through Design LightBlue' app and install it.

Open the app, give it permission to access the Bluetooth radio and your location if requested, and look through the list of nearby beacons for the one called `picow_ble`. You may find it helpful to use the app's filter option to narrow the list down to the strongest signals (a lower number indicates a stronger signal).

Click the **Connect** button and the LightBlue app will connect to Pico W over Bluetooth Low Energy. Go down to the **Environmental Sensing** section and tap on **Temperature** and you'll see your payload data — if not, tap the **Read Again** button to capture a fresh reading.

The temperature you receive, though, doesn't look like a number at first glance. That's because, by default, LightBlue decodes the structured data into hexadecimal. To make the number more easily readable, look for a drop-down box next to **Data format** which says **Hex**. Tap it, and choose **Signed Little-Endian** from the list: this will display the beacon data in decimal, or base ten — the normal counting system you use every day. Finally, you need to undo the multiplication your helper function does: divide the displayed number by 100 to get the reported temperature in degrees Celsius: a value of 3032, for example, means a temperature of 30.32°C.

Congratulations: your Pico W is now a wireless temperature sensor! If you want to be able to deploy your Pico W away from your computer as a remote sensor, remember to save a copy of your program as **main.py** — then it'll automatically run every time you connect Pico W to power, either via a USB power supply or a battery.

Receiving services

While you can use your smartphone or tablet to receive the beacon broadcasts from Pico W, you can also use another Pico W — if you have one. If you only have one Pico W, feel free to skip this part of the chapter — otherwise, make sure you've saved your temperature beacon program as **main.py** and disconnect your first Pico W, connect your second Pico W in its place, then open Thonny and start a new program:

```
import struct
import asyncio
import aioble
import bluetooth

ble_name = "picow_ble"
ble_svc_uuid = bluetooth.UUID(0x181A)
ble_characteristic_uuid = bluetooth.UUID(0x2A6E)
ble_scan_length = 5000
ble_interval = 30000
ble_window = 30000
```

Here you're using the same libraries as before, and some of the same variables — only this time, rather than setting the beacon name and UUIDs you're telling Pico W what it should be looking for when it scans for beacons.

There are three other variables here too: **ble_scan_length** sets the time, in milliseconds, that a scan for BLE advertising beacons should last before the program gives up, while **ble_interval** and **ble_window** configure detection timings in microseconds — setting them low in order to improve the likelihood of the two Pico W boards successfully seeing each other and connecting.

You're going to be creating asynchronous coroutines again, starting with one using **aioble** to scan for a BLE beacon which matches the details you set above — one with the name **picow_ble**, the service UUID of 0x181A, and characteristic ID of 0x2A6E. Type the following:

```
async def ble_scan():
    print("Scanning for BLE beacon named", ble_name, "...")
    async with aioble.scan(
    ble_scan_length,
    interval_us=ble_interval,
    window_us=ble_window,
    active=True) as scanner:
        async for result in scanner:
            if result.name() == ble_name and \
                ble_svc_uuid in result.services():
                    return result.device
    return None
```

This coroutine prints a message to the shell and begins a five-second scan for BLE beacons in the area. It then checks each beacon's name and service UUID against the variables you set earlier in the program — and if there's a match, returns the device's details for use later in the program.

Next, you'll need to create another helper function for handling structured data. Type the following:

```python
def decode_temp(data):
    return struct.unpack("<h", data)[0] / 100
```

If you compare this to the helper function in your earlier program, you'll notice it's the exact opposite: where your previous function converted the temperature reading into an integer, multiplied it by 100, and packed it as structured data, this one unpacks structured data and divides the resulting value by 100 — getting you back to a temperature reading in degrees Celsius.

Now your program needs to handle the output of the scan. Type the following:

```python
async def main():
    device = await ble_scan()
    if not device:
        print("BLE beacon not found.")
        return

    try:
        print("Connecting to", device)
        connection = await device.connect()
    except asyncio.TimeoutError:
        print("Connection timed out.")
        return
```

Here you're setting up some error handling, in case your scan can't find the BLE beacon running on your other Pico W. Assuming it does find it, your program will try to connect — handling another error if the connection fails.

To actually do something once connected, you'll need some more code. Add the following, paying careful attention to indentation and making sure the first line is indented by four spaces to make it part of the **main()** coroutine:

```python
    async with connection:
        try:
            ble_service = await connection.service(ble_svc_uuid)
            ble_characteristic = await \
              ble_service.characteristic(ble_characteristic_uuid)
        except (asyncio.TimeoutError, AttributeError):
            print("Timeout discovering services/characteristics.")
            return
```

```
while True:
    temp = decode_temp(await ble_characteristic.read())
    print("Temperature:", temp)
    await asyncio.sleep_ms(2000)
```

This, again, includes some error handling — this time in case the connection is successful, but the beacon doesn't send its service or characteristic UUIDs. This shouldn't happen, unless the beacon Pico W was unplugged just after connecting — or if the two Pico Ws are just slightly too far apart and suffering from a weak signal strength.

Assuming there's no error, the code then calls the helper function to take the beacon payload and decode it into a temperature value before printing it to the shell. Finally, there's a non-blocking delay — set to two seconds, matching the delay in your beacon program — before the next reading is printed to the shell.

As before, you'll need one all-important line to set things into motion:

```
asyncio.run(main())
```

Save this to your second Pico W as **BLE_Temperature_Client.py**. Click **Run**, and watch the shell area: you'll see the scan for beacons begin, then exit with an error after failing to find a match.

Power up your original Pico W, with the temperature beacon program on it as **main.py.** Give the beacon Pico W a second or two to set things up, then click Run again: this time your scanning Pico W should find the beacon and automatically connect.

Look in the shell area of the Thonny window, and you'll see the beacon content from your first Pico W printed out — after conversion from the structured data to a human-friendly temperature measurement in degrees Celsius. The second Pico W will *subscribe* to the beacon, meaning that it will continue to receive new data every time it's transmitted — printing the latest reading to the shell once every two seconds.

Congratulations: you've now connected two Pico Ws wirelessly over Bluetooth Low Energy, creating your own Internet of Things in miniature!

FURTHER READING

You can learn more about Pico W's Bluetooth capabilities and using them at a lower level, without the **aioble** library, in the free book *Connecting to the Internet with Raspberry Pi Pico W*, available from **rptl.io/picow-connect**.

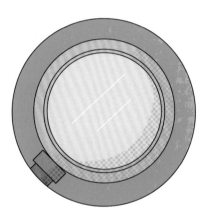

Appendix A

Raspberry Pi Pico specifications

The various components and features of a microcontroller are known as its *specifications*, and a look at the specifications gives you the information you need to compare two microcontrollers.

These specifications can seem confusing at first, are highly technical, and you don't need to know them to use your Raspberry Pi Pico, but they are included here for the curious reader.

Raspberry Pi Pico's microcontroller chip is a Raspberry Pi RP2040, which you'll see indicated by markings etched into the top of the component if you look closely enough. The microcontroller's name can be broken down into sections, each of which has a particular meaning:

- **RP** means 'Raspberry Pi', simply enough.

- **2** is the number of *processor cores* the microcontroller has.

- **0** is the type of processor core, indicating in this case the RP2040 uses a processor core called the *Cortex-M0+* from Cambridge-based Arm.

- **4** is how much *random access memory* (*RAM*) the microcontroller has, based on a special mathematical function: `floor(log2(RAM/16))`. In this case, '4' means the chip has *264 kilobytes* (*kB*) of RAM.

- **0** is how much *non-volatile* (*NV*) storage the chip has, and is worked out in the same way as the RAM: `floor(log2(NV/16))`. In this case, 0 simply means there is no non-volatile storage on-board.

The RP2040 is the first microcontroller from Raspberry Pi; when future models are released, these numbers will be used so you can quickly see how their features compare.

Your Pico's two Cortex-M0+ processor cores run at 48MHz (48 million cycles per second), though this can be changed in software up to 133MHz (133 million cycles per second) if your program needs higher performance.

The microcontroller's RAM is built into the same chip as the processor's cores themselves, and takes the form of six individual memory banks totalling 264kB (264,000 bytes) of static RAM (SRAM). The RAM is used to store your programs and the data they need.

RP2040 includes 30 multifunction general-purpose input/output (GPIO) pins, 26 of which are brought out to physical pin connectors on your Pico and one of which is connected to an on-board LED (on the Pico, that is; on the Pico W, the on-board LED is connected to a GPIO on the wireless module). Three of these GPIO pins are connected to an analogue-to-digital converter (ADC), while another ADC channel is connected to an on-chip temperature sensor.

RP2040 includes two *universal asynchronous receiver-transmitter* (*UART*), two *serial peripheral interface* (*SPI*), and two inter-integrated circuit (I2C) buses for connections to external hardware devices like sensors, displays, digital-to-analogue converters (DACs), and more. The microcontroller also includes *programmable input/output* (*PIO*), which lets the programmer define new hardware functions and buses in software.

Your Pico includes a micro USB connector, which provides a UART-over-USB serial link to the RP2040 microcontroller for programming and interaction, and which powers the chip. Hold down the BOOTSEL button when plugging the cable to switch the microcontroller into *USB Mass Storage Device mode*, allowing you to load new *firmware*.

RP2040 also includes an accurate *on-chip clock and timer*, which allows it to keep track of the time and date. The clock can store the year, month, day, day of the week, hour, minute, and second, and automatically keeps track of elapsed time as long as power is provided.

Finally, RP2040 includes *single-wire debug* (*SWD*) for hardware debugging purposes, brought out to three pins at the bottom of your Pico. **Figure A-1** shows the major on-board components of the Raspberry Pi Pico.

▸ **CPU:** 32-bit dual-core ARM Cortex-M0+ up to 133MHz (48MHz default)

▸ **RAM:** 264kB of SRAM in six independently configurable banks

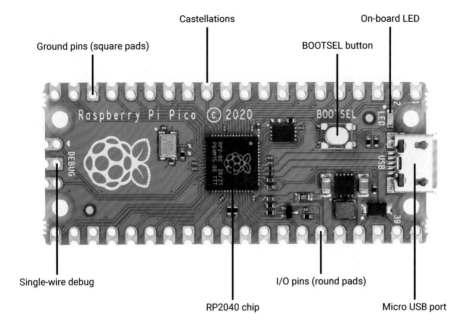

Ground pins (square pads)　Castellations　BOOTSEL button　On-board LED

Single-wire debug　RP2040 chip　I/O pins (round pads)　Micro USB port

Figure A-1　Raspberry Pi Pico

- ▶ **Storage:** 2MB external flash RAM

- ▶ **GPIO:** 26 pins

- ▶ **ADC:** 3 × 12-bit ADC pins

- ▶ **PWM:** Eight slices, two outputs per slice for 16 total

- ▶ **Clock:** Accurate on-chip clock and timer with year, month, day, day-of-week, hour, second, and automatic leap-year calculation

- ▶ **Sensors:** On-chip temperature sensor connected to 12-bit ADC channel

- ▶ **LEDs:** On-board user-addressable LED

- ▶ **Bus Connectivity:** 2 × UART, 2 × SPI, 2 × I2C, 2 × programmable I/O (PIO) blocks (8 state machines in total)

- ▶ **Hardware Debug:** Single-Wire Debug (SWD)

- ▶ **Mount Options:** Through-hole and castellated pins (unpopulated) with 4 × mounting holes

- ▶ **Power:** 5 V via micro USB connector, 3.3 V via 3V3 pin, or 2–5V via VSYS pin

The Raspberry Pi Pico W has the same specifications, but also adds the following, powered by an Infineon CYW43439:

- **Wi-Fi:** Single-band 802.11n on a 2.4GHz wireless interface

- **Personal Area Network:** Bluetooth 5.2, supporting Bluetooth LE Central and Peripheral roles, as well as Bluetooth Classic.

- **Antenna:** onboard antenna licensed from ABRACON (formerly ProAnt)

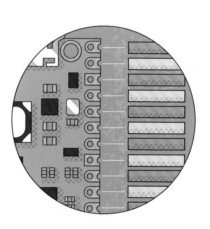

Appendix B

Pinout guide

Raspberry Pi Pico exposes 26 of the 30 possible RP2040 GPIO (general-purpose input/output) pins by routing them straight out to Pico header pins. GP0 to GP22 are digital only and GP 26–28 are able to be used either as digital GPIO or as ADC (analogue-to-digital converter) inputs, selectable in software. Most of the GPIO pins also offer secondary functionality for SPI, I2C, or UART communication protocols. All GPIO pins may also be used with PWM (pulse-width modulation) — see Chapter 8, *Temperature gauge* for more details.

- ▸ **VBUS** is the micro USB input voltage, connected to micro USB port pin 1. This is nominally 5 V (0 V if the USB is not connected or not powered).

- ▸ **VSYS** is the main system input voltage, which can vary in the allowed range 1.8 V to 5.5 V, and which is used by the on-board SMPS (switch mode power supply) to generate the 3.3 V for the RP2040 and its GPIO.

- ▸ **3V3_EN** connects to the on-board SMPS enable pin, and is pulled high (to VSYS) via a 100 kΩ resistor. To disable the 3.3 V (which also de-powers the RP2040), short this pin low.

- ▸ **3V3** is the main 3.3 V supply to RP2040 and its IO, generated by the on-board SMPS. This pin can be used to power external circuitry (the maximum output current will depend on the RP2040 load and VSYS voltage, but you must keep the load on this pin less than 300 mA).

- ▸ **ADC_VREF** is the ADC power supply (and reference) voltage, and is generated on Pico by filtering the 3.3 V supply. This pin can be used with an external reference if better ADC performance is required.

- ▸ **AGND** is the ground reference for GPIO26–29; there is a separate analogue ground plane running under these signals and terminating at this pin. If the ADC is not used or ADC performance is not critical, this pin can be connected to digital ground.

‣ **RUN** is the RP2040 enable pin, and has an internal (on-chip) pull-up resistor to 3.3 V of about ~50 kΩ. To reset RP2040, short this pin low.

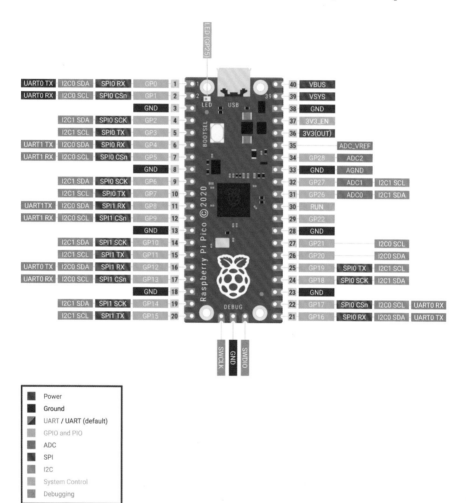

Figure B-1   Raspberry Pi Pico pinout

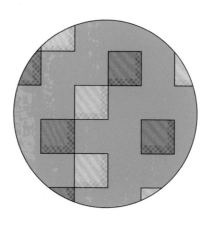

Appendix C

Programmable I/O

In this appendix, we look at code that looks very different from the code we've dealt with in the rest of the book. That's because we'll be dealing with things at a low level. Most of the time, MicroPython hides the complexities of how things work on the microcontroller. When we do something like:

```
print("hello")
```

...we don't have to worry about the way the microcontroller stores the letters, or the format in which they get sent to the serial terminal, or the number of clock cycles the serial terminal takes. This is all handled in the background. However, when we get to Programmable Input and Output (PIO), we need to deal with things at a much lower level.

We're going to go on a whistle-stop tour of PIO and introduce some advanced topics so you can get an idea of what's going on, and hopefully understand how PIO on Pico offers some real advantages over the options you'll find on other microcontrollers.

However, understanding all the low-level data manipulation required to create PIO programs takes time to fully get your head around, so don't worry if it seems a little opaque. If you're interested in tackling this low-level programming, then we'll give you the knowledge to get started and point you in the right direction to continue your journey. If you're more interested in working at a higher level and would rather leave the low-level wrangling to others, we'll show you how to *use* PIO programs.

Data in and data out

Throughout this book, we've looked at ways of controlling the pins on your Pico using MicroPython. We can switch them on and off, take inputs, and even send data using the dedicated SPI and I2C controllers. However, what if we want to connect a device that doesn't communicate in SPI or I2C? What about a device with its own special protocol?

There are a couple of ways we can do this. On most MicroPython devices, you need to do a process called *bit banging* where you implement the protocol in MicroPython. Using this, you turn the pins on or off in the right order to send data.

There are three downsides to this. The first is that it's slow. MicroPython does some things really well, but it doesn't run as fast as natively compiled code.

The second is that we have to juggle this with the rest of our code that is running on the microcontroller.

The third is that some timing-critical code can be hard to implement reliably. Fast protocols often require that things happen at very precise times, and with MicroPython we can be quite precise, but if you're trying to transfer megabits a second, you need things to happen every millisecond or possibly every few hundred nanoseconds. That's hard to do (reliably) in MicroPython.

Pico has a solution to this: *Programmable I/O*. There are some extra, really stripped-back processing cores that can run simple programs to control the I/O pins. You can't program these cores *with* MicroPython — you must use a special language just for them — but you can program them *from* MicroPython. Here's an example:

```
from rp2 import PIO, StateMachine, asm_pio
from machine import Pin
import time

@asm_pio(set_init=PIO.OUT_LOW)
def quarter_bright():
    set(pins, 0) [2]
    set(pins, 1)

@asm_pio(set_init=PIO.OUT_LOW)
def half_bright():
    set(pins, 0)
    set(pins, 1)

@asm_pio(set_init=PIO.OUT_HIGH)
```

```
def full_bright():
    set(pins, 1)

led = Pin("LED")
sm1 = StateMachine(1, quarter_bright, freq=10000, set_base=led)
sm2 = StateMachine(2, half_bright, freq=10000, set_base=led)
sm3 = StateMachine(3, full_bright, freq=10000, set_base=led)

while(True):
    sm1.active(1)
    time.sleep(1)
    sm1.active(0)

    sm2.active(1)
    time.sleep(1)
    sm2.active(0)

    sm3.active(1)
    time.sleep(1)
    sm3.active(0)
```

There are three methods here that all look a little strange; they set the on-board LED to quarter, half, and full brightness. The reason they look a little strange is because they're written in a special language for the PIO system of Pico. You can probably guess what they do — flick the LED on and off very quickly in a similar way to how we used PWM. The instruction **set(pins, 0)** turns a GPIO pin off and **set(pins, 1)** turns the GPIO pin on.

Each of the three methods has a descriptor above it that tells MicroPython to treat it as a PIO program and not a normal method. These descriptors can also take parameters that influence the behaviour of the programs. In these cases, we've used the **set_init** parameter to tell the PIO whether the GPIO pin should start off being low or high.

Each of these methods — which are really mini programs that run on the PIO state machines — loops continuously. So, for example, **half_bright** will constantly turn the LED on and off so that it spends half its time off and half its time on. **full_bright** will similarly loop, but since the only instruction is to turn the LED on, this doesn't actually change anything.

The slightly unusual one here is **quarter_bright**. Each PIO instruction takes exactly one clock cycle to run (the length of a clock cycle can be changed by setting the frequency, as we'll see later). However, we can add a number between 1 and 31 in square brackets after an instruction, and this tells the PIO state machine to delay execution by this number of clock cycles before running the next instruction. In **quarter_bright**, then, the two **set** instructions each take one clock cycle, and the delay takes two clock cycles, so the total loop takes four clock cycles. In the first line, the **set** instruction takes one cycle and the delay takes two, so the GPIO pin is off for three of these four cycles. This makes the LED a quarter as bright as if it were on constantly.

Once you've got your PIO program, you need to load it into a *state machine*. Since we have three programs, we need to load them into three different state machines (there are eight you can use, numbered 0–7). You can load a PIO program with a line like this:

```
sm1 = StateMachine(1, quarter_bright, freq=10000, set_base=led)
```

The parameters here are:

- ▶ The state machine number

- ▶ The PIO program to load

- ▶ The frequency (which must be between 2000 and 125000000)

- ▶ The GPIO pin that the state machine manipulates

There are some additional parameters you'll see in other programs that we don't need here.

Once you've created your state machine, you can start and stop it using the **active** method with 1 (to start) or 0 (to stop). In our loop, we cycle through the three different state machines.

A real example

The previous example was a little contrived, so let's look at a way of using PIO with a real example. WS2812B LEDs (sometimes known as NeoPixels) are a type of light that contains three LEDs (one red, one green, and one blue) and a small microcontroller. They're controlled by a single data wire with a timing-dependent protocol that's hard to bit-bang.

Wiring your LED strip is simple, as shown in **Figure C-1**. Depending on the manufacturer of your LED strip, you may have the wires already connected, you may have a socket that you can push header wires in, or you may need to solder them on yourself.

One thing you need to be aware of is the potential current draw. While you can add an almost endless series of NeoPixels to your Pico, there's a limit to how much power you can get out of the 5 V pin on Pico. Here, we'll use eight

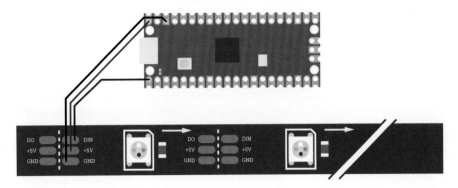

Figure C-1 Connecting an LED strip

LEDs, which is perfectly safe, but if you want to use many more than this, you need to understand the limitations and may need to add a separate power supply. You can cut a longer strip to length, and there should be cut lines between the LEDs to show you where to cut. There's a good discussion of the various issues at **hsmag.cc/neopixelpower**.

Now we've got the LEDs wired up, let's look at how to control it with PIO:

```python
import array, time
from machine import Pin
import rp2
from rp2 import PIO, StateMachine, asm_pio

# Configure the number of WS2812 LEDs.
NUM_LEDS = 8

@asm_pio(sideset_init=PIO.OUT_LOW, out_shiftdir=PIO.SHIFT_LEFT,
         autopull=True, pull_thresh=24)
def ws2812():
    T1 = 2
    T2 = 5
    T3 = 3
    label("bitloop")
    out(x, 1)               .side(0) [T3 - 1]
    jmp(not_x, "do_zero")   .side(1) [T1 - 1]
    jmp("bitloop")          .side(1) [T2 - 1]
    label("do_zero")
    nop()                   .side(0) [T2 - 1]

# Create a StateMachine with the ws2812 code and output on Pin(0).
```

```
sm = StateMachine(0, ws2812, freq=8000000, sideset_base=Pin(0))

# Start the StateMachine, it will wait for data on its FIFO.
sm.active(1)
```

The basic way that this works is that 800,000 bits of data are sent per second (notice that the frequency is 8000000 and each cycle of the program is 10 clock cycles). Every bit of data is a pulse — a short pulse indicating a 0 and a long pulse indicating a 1. A big difference between this and our previous program is that MicroPython needs to be able to send data to this PIO program.

There are two stages for data coming into the state machine. The first is a bit of memory called a First In, First Out queue (or FIFO). This is what our main Python program sends data *to*. The second is the Output Shift Register (OSR). This is where the **out()** instruction fetches data *from*. The two are linked by *pull instructions* which take data from the FIFO and put it in the OSR. However, since our program is set up with **autopull** enabled with a threshold of 24, each time we've read 24 bits from the OSR, it will be reloaded from the FIFO.

The instruction **out(x,1)** takes one bit of data from the OSR and places it in a variable called **x** (there are only two available variables in PIO: **x** and **y**).

The **jmp** instruction tells the code to move directly to a particular label, but it can have a condition. The instruction **jmp(not_x, "do_zero")** tells the code to move to **do_zero** if the value of **x** is 0 (or, in logical terms, if **not_x** is true, and **not_x** is the opposite of **x** — in PIO-level speak, 0 is false and any other number is true).

There's a bit of **jmp** spaghetti that is mostly there to ensure that the timings are consistent because the loop has to take exactly the same number of cycles every iteration to keep the timing of the protocol in line.

The one aspect we've been ignoring here is the **.side()** bits. These are similar to **set()** but they take place at the same time as another instruction. This means that **out(x,1)** takes place as **.side(0)** is setting the value of the sideset pin to 0.

Phew, that's quite a bit going on for such a small program. Now we've got it active, let's look at how to use it. Add the following MicroPython code under the preceding code to send data to a PIO program.

```
# Display a pattern on the LEDs via an array of LED RGB values.
ar = array.array("I", [0 for _ in range(NUM_LEDS)])

print("blue")
```

```
for j in range(0, 255):
    for i in range(NUM_LEDS):
        ar[i] = j
    sm.put(ar,8)
    time.sleep_ms(10)

print("red")
for j in range(0, 255):
    for i in range(NUM_LEDS):
        ar[i] = j<<8
    sm.put(ar,8)
    time.sleep_ms(10)

print("green")
for j in range(0, 255):
    for i in range(NUM_LEDS):
        ar[i] = j<<16
sm.put(ar,8)
time.sleep_ms(10)

print("white")
for j in range(0, 255):
    for i in range(NUM_LEDS):
        ar[i] = (j<<16) + (j<<8)
    sm.put(ar,8)
    time.sleep_ms(10)
```

Here we keep track of an array called **ar** that holds the data we want our LEDs to have (we'll look at why we created the array this way in a little while). Each number in the array contains the data for all three colours on a single light. The format is a little strange as it's in binary. One thing about working with PIO is that you often need to work with individual bits of data. Each bit of data is a 1 or 0, and numbers can be built up in this way, so the number 2 in base 10 (as we call the normal numbers we're used to using) is 10 in binary. 3 in base 10 is 11 in binary. The largest number in eight bits of binary is 11111111, or 255 in base 10. We won't go too deep into binary here, but if you want to find out more, you can try the Binary Hero project here: **hsmag.cc/binaryhero**.

To make matters a little more confusing, we're actually storing three numbers in a single number. This is because in MicroPython, whole numbers are stored in 32 bits, but we only need eight bits for each number. There's a little free space at the end as we really only need 24 bits, but that's OK.

The first eight bits are the blue values, the next eight bits are red, and the final eight bits are green. The maximum number you can store in eight bits is 255, so each LED has 255 levels of brightness. We can do this using the bit shift operator `<<`. This adds a certain number of 0s to the end of a number, so if we want our LED to be at level 1 brightness in red, green, and blue, we start with each value being 1, then shift them the appropriate number of bits. For green, we have:

```
1 << 16 = 10000000000000000
```

For red we have:

```
1 << 8 = 100000000
```

And for blue, we don't need to shift the bits at all, so we just have 1. If we add all these together, we get the following (if we add the preceding bits to make it a 24-bit number):

```
000000010000000100000001
```

The rightmost eight bits are the blue, the next eight bits are red, and the left-most eight bits are green. The part that may seem a bit confusing is this line from the top of the code you just added:

```
ar = array.array("I", [0 for _ in range(NUM_LEDS)])
```

This creates an array which has **I** as the first value, and then a 0 for every LED. The reason there's an **I** at the start is that it tells MicroPython that we're using a series of 32-bit values. However, we only want 24 bits of this sent to the PIO for each value, so we tell the **put** command to remove eight bits with:

```
sm.put(ar,8)
```

All the instructions

The language used for PIO state machines is very sparse, so there are only a small number of instructions. In addition to the ones we've looked at, you can use:

- ▸ **in()** — moves between 1 and 32 bits into the state machine (similar, but opposite to **out()**).

- ▸ **push()** — sends data to the memory that links the state machine and the main MicroPython program.

- ▸ **pull()** — gets data from the chunk of memory that links the state machine and the main MicroPython program. We haven't used it here, because by including `autopull=True` in our program, this happens automatically when we use `out()`.

▸ `mov()` — moves data between two locations (such as between the **x** and **y** variables or a register like OSR).

▸ `irq()` — controls interrupts. These are used if you need to trigger a particular thing to run on the MicroPython side of your program.

▸ `wait()` — pauses until something happens (such as a I/O pin changes to a set value or an interrupt happens).

WS2812B LIBRARY

While it's useful to experiment with the WS2812B PIO program, if you want to use it in a real project, it may be more useful to use a library that brings it all together. There's one such example at **hsmag.cc/pico-ws2812b**. This lets you create an object that holds all the LED colour data and then use methods such as `set_pixel()` and `fill()` to alter the data. Look in the examples folder of that repository for more details of how to use it.

Although there are only a small number of possible instructions, it's possible to implement a huge range of communications protocols. Most of the instructions are for moving data about in some form. If you need to prepare the data in any particular way, such as manipulating the colours you want your LEDs to be, this should be done in your main MicroPython program rather than the PIO program.

You can find more information on how to use these, and the full range of options for PIO in MicroPython, on Raspberry Pi Pico in the Pico Python SDK document — and a complete reference to how PIO works in the RP2040 databook. Both of these are available at **rptl.io/rp2040-get-started**.